Supported by the National Natural Science Foundation of China (No.40930315)

Supported by the National Fund for Academic Publication in Science and Technology

Early Permian Tarim Large Igneous Province in Northwest China

Shufeng Yang, Hanlin Chen, Zilong Li, Yinqi Li, Xing Yu

图书在版编目（CIP）数据

塔里木早二叠世大火成岩省 = Early Permian Tarim Large Igneous Province in Northwest China ：英文 / 杨树锋等著. —杭州：浙江大学出版社，2017.12
ISBN 978-7-308-17664-4

Ⅰ.①塔… Ⅱ.①杨… Ⅲ.①塔里木盆地－早二叠世－火成岩－研究－英文 Ⅳ.①P588.1

中国版本图书馆CIP数据核字(2017)第283663号

塔里木早二叠世大火成岩省

杨树锋 陈汉林 厉子龙 励音骐 余 星 著

丛书统筹	国家自然科学基金委员会科学传播中心
策划编辑	徐有智 许佳颖
责任编辑	伍秀芳 (wxfwt@zju.edu.cn)
责任校对	魏钊凌 董 唯
封面设计	程 晨
出版发行	浙江大学出版社
	（杭州市天目山路148号 邮政编码310007）
	（网址：http://www.zjupress.com）
排 版	杭州中大图文设计有限公司
印 刷	浙江海虹彩色印务有限公司
开 本	710mm×1000mm 1/16
印 张	13.25
字 数	330
版 印 次	2017年12月第1版 2017年12月第1次印刷
书 号	ISBN 978-7-308-17664-4
定 价	128.00元

版权所有 翻印必究 印装差错 负责调换

浙江大学出版社发行中心联系方式（0571）88925591; http://zjdxcbs.tmall.com

Advances in China's Basic Research Editorial Board

Editor-in-Chief   YANG Wei

Deputy Editors   GAO Wen GAO Ruiping

Members

HAN Yu	WANG Changrui	ZHENG Yonghe
ZHENG Zhongwen	FENG Feng	ZHOU Yanze
GAO Tiyu	ZHU Weitong	MENG Qingguo
CHEN Yongjun	DU Shengming	WANG Qidong
LI Ming	QIN Yuwen	GAO Ziyou
DONG Erdan	HAN Zhiyong	YANG Xinquan
REN Shengli		

Preface to the Series

As Lao Tzu said, "A huge tree grows from a tiny seedling; a nine-storey tower rises up from a mound of earth." Basic research is the fundamental approach to foster innovation-driven development, and its level becomes an important yardstick for measuring the overall scientific and national strength of a country. Since the beginning of the 21st century, China's overall strength in basic research has been increasing consistently. With respect to input and output, China's input in basic research has increased by 14.8 times from 5.22 billion *yuan* in 2001 to 82.29 billion *yuan* in 2016, with an average annual increase of 20.2%. In the same period, the number of China's scientific papers included in *Science Citation Index* (SCI) increased from less than 40,000 to 324,000; China rose from the 6th place to the 2nd place in global ranking in terms of the number of published papers. In regard to the quality of output, in 2016, China ranked No.2 in the world in terms of citation in 9 disciplines, among which Materials Science ranked No.1; in the past two years, China ranked No.3 in the world both in the number of top 1% most-cited international papers and the number of top 1‰ international hot papers with global share of 25.1%. In talent cultivation, in 2016, 175 scientists from China were included in the Thomson Reuters Highly-Cited Researchers List (136 of which from the Chinese Mainland), which ranked the fourth in the world and the first in Asia.

Meanwhile, we should also be keenly aware that China's basic research is still subject to great challenges. First, funding for basic research in China is still far less than that in developed countries — only about 5% of the R&D funds in China are used for basic research, a much lower percentage than the 15%–20% in developed countries. Second, competence for original innovation in China is insufficient. The major original science achievements that have global impact are still rare. Most of the scientific research projects are just a follow-up and imitation of the existing researches, rather than brand new novel or pioneering work. Third, the development of disciplines is not balanced, and China's research level in some disciplines is noticeably lower than the international level — China's Field-Weighted Citation Impact (FWCI) in disciplines just reached 0.94 in 2016, lower than the world average of 1.0.

The Chinese government attaches great importance to basic research. In the "13th Five-Year Plan", China has confirmed scientific and technological innovation as a priority in all-round innovation, and has made strategic arrangements to strengthen basic research. General Secretary Xi Jinping put forward a grand blueprint of making China into a strong power in science and technology in his speech delivered at the National Conference on Scientific and Technological Innovation, and placed emphases on "targeting the world's advanced scientific and technological frontier, consolidating basic research to achieve major breakthroughs in forward-looking basic research and steering original achievements" at the 19th CPC National Congress on Oct.18, 2017. With more than 30 years of unremitting exploration, the National Natural Science Foundation of China (NSFC), one of the main channels for supporting basic research in China, has gradually shaped a funding pattern covering research, talents, tools and convergence, and has taken actions to vigorously promote basic frontier research and the growth of scientific research talents, reinforce the building of innovative research teams, deepen regional cooperation and exchanges, and push forward multidisciplinary convergence. As of 2016, nearly 70% of China's published scientific papers were funded by NSFC — accounted for 1/9 of the total number of published papers all over the world. Facing the new strategic target of building China into a strong country in science and technology, NSFC will conscientiously reinforce forward-looking planning, and enhance the efficiency of evaluation, so as to achieve the strategic goal of making China progressively share the same level with major innovative countries in research output, impact and original contribution by 2050.

The series of *Advances in China's Basic Research* and the series of *Reports of China's Basic Research* proposed and planned by NSFC emerge under such a background. Featuring of science, basics and advances, the two series are aimed to share innovative achievements, diffuse performances of basic research, and lead breakthroughs in key fields. They will closely follow the frontiers of basic research developments in China, and publish excellent innovation achievements funded by NSFC. The series of *Advances in China's Basic Research* will mainly present the important original achievements of the programs funded by NSFC and display the breakthroughs and forward guidance of the key research fields, while the series of *Reports of China's Basic Research* will show the core contents of the final reports of the Major Programs and the Major Research Plans funded by NSFC to make a systematical summarization and strategic outlook of the achievements in the fields preferred to be funded by NSFC. We not only hope to comprehensively and systematically display the backgrounds, scientific significances, discipline layouts, frontier breakthroughs of the programs, as well as strategic outlooks of the subsequent research, but also expect to summarize the innovative ideas, enhance multidisciplinary convergence and promote the continuity of research in the fields concerned as well as original discoveries.

As an old saying in *Hsun Tzu* goes, "Where accumulated earth becomes a mountain, there prevails wind and rain. Where running waters gather widely and deeply, there gives birth to dragons." The series of *Advances in China's Basic Research* and the series of *Reports of China's Basic Research* are hoped to become the "historical records" of China's basic research, which will provide researchers with abundant scientific research materials and sources for innovation. It's believed that the series will certainly play an active role in making China's basic research prosper and in building China into a powerful nation of science and technology.

President of NSFC
Academician of Chinese Academy of Sciences
Dec. 2017, Beijing

Preface

Tarim Plate is one of the three major plates in China, surrounded by the Tianshan, Kunlun and Altyn–Tagh orogenic belts. It is also an important connection to the tectonic domains between Central Asia and the Tethys. One notable feature of the Tarim Plate is the wide occurrence of Early Permian intraplate magmatism in which the magmatic rocks were made up mainly of basaltic rocks including basalts, diabase, basaltic andesite, ultramafic rocks, etc. The area of residual distribution of the magmatic rocks can reach about 2.5×10^5 km^2, and the largest residual thickness is more than 700 m. As a new Large Igneous Province (LIP), it has attracted the attention of many scientists.

Based on more than 20 years of study, this book will systematically introduce the tempo-spatial features of the Early Permian Tarim Large Igneous Province (Tarim LIP), the geochemical features and the magma evolution of the rock units, as well as the geodynamics and metallogenesis of the Early Permian Tarim LIP. This book will also provide a new geodynamic model for the LIPs, which is different from the model based on the Deccan LIP and the Parana LIP. This book is the first book to introduce the Early Permian Tarim LIP, and it is an ideal book for researchers and graduate students in tectonics, igneous petrology, geochemistry, geophysics, Earth evolution and planetary geology, as well as for professionals working in the mining industry. This book will also play a very important role in the study of the LIPs and geodynamic research within the tectonic domain of Central Asia. By taking this opportunity, we would like to express our sincere appreciation of the assistance from other teachers and students at the School of Earth Sciences at Zhejiang University. The research work in this book was supported jointly by the National Natural Science Foundation of China (Nos. 41603029 and 40930315), the National Science and Technology Major Project of China (No. 2017ZX005008-001), the National Key R&D Program of China (No. 2016YFC0601004), and the National Basic Research Program of China (Nos. 2007CB411303 and 2011CB808902).

The Authors
October 2017, Hangzhou

Contents

Chapter 1 Introduction 1
  1.1 Tectonic Evolution of the Tarim Block 1
  1.2 Brief Introduction to LIPs 10
  1.3 Research History on the Early Permian Tarim Large Igneous Province (Tarim LIP) 13
  References 18

Chapter 2 Tempo-Spatial Features of the Tarim LIP 27
  2.1 Lithological Characteristics of Different Rock Units in the Tarim LIP 28
  2.2 Spatial Distribution of the Tarim LIP 37
  2.3 Time Sequence of the Tarim LIP 53
  References 65

Chapter 3 Geochemical Features of the Tarim LIP Rocks and Implications for the Magma Evolution 75
  3.1 General Geochemical Features of the Tarim LIP Rocks 76
  3.2 Geochemical Features of the Three Basalt Groups 78
  3.3 Geochemical Features of the Bachu Intrusive Rocks 82
  3.4 Effects of Crustal Contamination on the Tarim CFBs 84

3.5 Implications for Source Isotopic Heterogeneity and
Plume–Lithosphere Interaction in the Tarim LIP 90

References 92

Supplementary Table I 96

Supplementary Table I's References 106

Chapter 4 Geodynamics of the Tarim LIP 109
4.1 Relationship Between the Tarim LIP and the Mantle Plume ... 109
4.2 Geochemical Comparison with other Permian Magmatism in
Central Asia 120
4.3 Geodynamic Model of the Tarim LIP 135
References 144

Chapter 5 Metallogenesis of the Tarim LIP 153
5.1 Wajilitag Fe–Ti–V Oxide Deposit 154
5.2 Cu–Ni–PGE Mineral Resource Potential in the Tarim LIP 165
References 175

Supplementary Table II 181

Supplementary Table III 183

Supplementary Table IV 189

Supplementary Table V 191

Supplementary Table VI 193

Index 195

1

Introduction

Abstract: This chapter includes the tectonic evolution of the Tarim Block, a brief introduction to Large Igneous Provinces (LIPs) and a history of the research on the Early Permian Tarim Large Igneous Province (Tarim LIP). The first part introduces the evolution of the basement during the tectonic episodes at 2.65–2.45 Ga, 2.0–1.8 Ga and 1.1–0.9 Ga, and the evolution of sedimentary cover sequences since the Late Neoproterozoic and igneous activity. The second part introduces the concept of LIPs, their classification, the age spectrum of selected LIP events through time and the effect of LIPs on the evolution of the Earth, environmental change and ore deposits, etc. The third part introduces a history of research on and the basic characteristics of the Tarim LIP.

Keywords: Tectonic evolution; History of research; Early Permian Tarim Large Igneous Province (Tarim LIP); Tarim Block

1.1 Tectonic Evolution of the Tarim Block

The Tarim Block and the North and South China Blocks make up the three major continental blocks in China. The Tarim Block occurs within the Xinjiang Uygur Autonomous Region of northwestern China and covers an area of more than 6×10^5 km². It is surrounded by the Tianshan orogen to the north, the Kunlun orogen to the south, and the Altyn-Tagh orogen to the southeast (Fig. 1.1). The main part of the Tarim Block is the Tarim Basin. The Tarim Basin can be divided into several tectonic units including the Kuqa Depression, the North Uplift, the North Depression, the Central Uplift, the Southwest Depression, the Southeast Uplift and the Southeast Depression (Fig. 1.1 and Fig. 1.2). A series of important tectonic movements occurred within superimposed basins during different periods. The features and textures of proto-type basins were generally superimposed and reconstructed by tectonic movements in later periods due to the unstable tectonics of the Tarim Block which resulted from the relatively smaller scale of the Tarim Craton Block and multiple, intense episodes of tectonic movements in peripheral areas (Jia and Wei, 2002; Jia et al., 2004).

The Tarim Block is a cratonic block with Archean and Paleo- to Meso-Proterozoic crystalline basements. Sedimentary cover is composed of Neo-Proterozoic, Paleozoic, Mesozoic and Cenozoic.

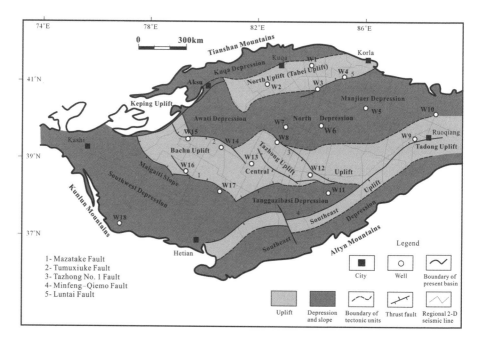

Fig. 1.1 Schematic tectonic map of the Tarim Basin, showing the distribution of tectonic units within the basin (After Liu et al., 2012)

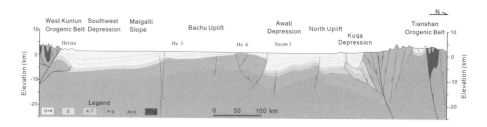

Fig. 1.2 The North to South profile of the Tarim Basin

Most of the Tarim Block is occupied by desert, but outcrops of Precambrian and Paleozoic to Cenozoic rocks are scattered along its margins. The Tarim Block is characterized by a double-layer structure consisting of a metamorphic basement overlain by the Late Neoproterozoic to Cenozoic sedimentary cover sequences.

1.1.1 Evolution of the Basement

The Tarim Block has experienced several stages of tectonic evolution since its formation, and the previous geochronological data from the TTG (Trondhjemite-Tonalite-Granodiorite) and other rocks from the Quruqtagh, Altyn-Tagh and Tiekelike areas suggest that the Tarim Block was mainly built up in several tectonic episodes at 2.65–2.45 Ga, 2.0–1.8 Ga and 1.1–0.9 Ga (Hu et al., 2000; Lu and Yuan, 2003; Zhu et al., 2008; Lu et al., 2008).

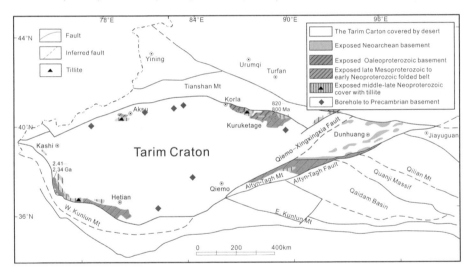

Fig. 1.3 Geological map showing the spatial distribution of Precambrian rocks in the Tarim Block (Revised from Zhao and Cawood, 2012; Xu et al., 2014)

The metamorphic basement of the Tarim Block is mainly composed of Archean to Early Neoproterozoic metamorphosed strata and magmatic rocks and it mainly crops out in four areas surrounding the orogenic belts of the basin, i.e. the Korla–Quruqtagh area, the Aksu–Keping area, the Tiekelike area and the Altun–Dunhuang areas at the NE, NW, SW and SE margins of the Tarim Block, respectively (Fig. 1.3). The basement consists of Neoarchean TTG gneisses with minor supracrustals, Paleoproterozoic mafic-felsic intrusions, high-grade supracrustals and minor anatectic granites, and Late Mesoproterozoic to Early–Middle Neoproterozoic meta-sedimentary and volcanic strata metamorphosed in greenschist and blueschist facies, which are together unconformably overlain by Late Neoproterozoic Sinian unmetamorphosed cover. This formation and evolution were closely related to the assembly and breakup of the supercontinents of Columbia (Nuna) and Rodinia (Lu et al., 2008; Zhang et al., 2012).

Neoarchean to Paleoproterozoic rocks in the Tarim Block mostly outcrop along its eastern and northern margins which are mainly exposed in the Quruqtagh and Dunhuang complexes, and include the Neoarchean tonalitic granitic rocks and the Paleoproterozoic amphibolite to granulite facies paragneiss, most of which were emplaced in the period 2.60–2.50 Ga (Lu, 1992; Long et al., 2010, 2011; Shu et al., 2011; Zhao and Gawood, 2012; Zhang et al., 2012). In most places, the Archean rocks outcrop as stripes or lenses with variable dimensions that are tectonically enclosed within the Paleoproterozoic paragneiss; both of them generally show foliations that are concurrent to each other (Zhang et al., 2012). These features suggest that the Archean and Paleoproterozoic rocks had undergone the same tectono–metamorphic event in the Paleoproterozoic Era because the low-grade metamorphic Mesoproterozoic unconformably overlies on the Archean and the Paleoproterozoic rocks (Xijiang BGMR, 1993). In the Qulukatage Complex, these Neoarchean and Paleoproterozoic rocks underwent two metamorphic events at 1.9–1.8 Ga and 1.1–1.0 Ga, which are considered as having been related to the assembly of the Columbia and Rodinia supercontinents, respectively (Shu et al., 2011; Zhang et al., 2012).

Late Mesoproterozoic to Early–Middle Neoproterozoic metamorphosed strata are exposed on the peripheral margins of the Tarim Block, represented by the Bowamu, Aierjigan and Aksu groups on the northern margin, the Kalakashi (Sailajiajitage) and Ailiankate groups on the southern margin, the Bulunkule Group on the southwestern margin, and the Altyn-Tagh Group on the southeastern margin. Most of these groups are considered to have formed in Andean-type continental margins, which were deformed and metamorphosed at 1.0–0.9 Ga, probably related to the assembly of Rodinia (Zhang et al., 2003; Lu et al., 2008).

1.1.2 Evolution of Sedimentary Cover Sequences

Since the late Middle Neoproterozoic, the Tarim Block has become a stable platform overlain by late Middle Neoproterozoic to Cambrian unmetamorphosed cover sequences (Fig. 1.4). The late Middle to Late Neoproterozoic sequences are called the Nahua and Sinian System containing four sequences of tillite, interpreted as evidence for the Neoproterozoic Snow Ball Earth Event. During the Nahua to Sinian, the Tarim Block began to break up during separation from the Rodinia supercontinent. Rifting-related mafic igneous rocks are widely distributed both in the northern and southern margins of the Tarim Block. The Nahua and Sinian sequences were deposited on the Pre-Nahua crystallized basement unconformably and are composed of glacial deposit conglomerate and terrigenous clastic deposits. The Sinian, overlying unconformably on the Nahua, consists mostly of dolomite and mudstone intercalated with basalt. The thickest deposits of the Sinian in the basin are found in the Manjiaer depression, which was controlled by a group of faults and filled with about 1000 m of deep marine mudstone and muddy limestone with rift volcanic rocks.

The Paleozoic series in Tarim exhibit typical features of the sedimentation formed at passive continental margins (Fig. 1.4). Along the southern–southeastern side and the northern side, the thickness of the Paleozoic reaches up to 12 km, and in central Tarim the thickness varies from 5000 m to 8000 m according to data from the boreholes (Jia et al., 2004). The thickness of the Paleozoic sequences in Tarim was strictly constrained by the sedimentary troughs between the Tarim and the neighboring orogens, i.e., the Tianshan orogen in the north and the western Kunlun–Altyn-Tagh orogen in the south and southeast. The large-scale transgression took place during the Early Cambrian and deposited a set of deep-water dark mudstones with phosphates at the base of the Cambrian over most parts of the block, which comprises one of the most important hydrocarbon source rocks in the basin. These pass upward into thin-bedded dolomites, thick gypsum and salt layers intercalated with dolomites (from the Middle Cambrian), formed in an evaporative carbonate platform and slope environment. The Upper Cambrian is made up of thick-bedded dolostone, and the Lower to Middle Ordovician consists mainly of thick dolomitic limestones and the low part of the Upper Ordovician is mainly made up of carbonate deposits formed in open platform environments with reef and shoal deposits along the platform margins. These deposits are overlain by silicic clastic abyssal deposits of the upper part of the Upper Ordovician.

System	Series	Formation	Age (Ma)	Seismic reflection surface	Lithology	Thickness (m)	Sequence 2nd order	Sequence 1st order	Depositoma; evolution and basin setting
Quaternary	Holocene		0.01			0 – 50		VIII	Alluvial and fluvial deposits
	Pleistocene	Xiyu	1.64	T₂		50 – 1336			Angular unconformity
Neogene	Pliocene	Kuqa	5.2	T₃		300 – 700	VII₃	VII	Alluvial, fluvial and shallow lacustrine deposits, foreland basin
		Kangeun	16.3	T₅		300 – 800	VII₃		
	Miocene	Jidike	23.3	T₆		600 – 800	VII₂		Parallel to angular unconformity
Paleogene	Oligocene	Suweiyi	35.4	T₇		200 – 400	VII₁		Alluvial, deltaic and lacustrine or lagoon deposits, forelang depression
	Eocene	Xikuzibai	56.5			400 – 600			
	Paleocene	Talake	65	T₈					Parallel to angular unconformity
Cretaceous	K₂	Bashenjiqike	91			100 – 300			Terrigenous deposits of inland depression or foreland marginal depression, dominated by alluvial, fluvial and shallow lacustrine deposits
		Baxigai	95				VI₃		
	K₁	Shushanhe				400 – 1060			
		Yageliemu	135	T₁₂		60 – 100			Parallel to angular unconformity
Jurassic	J₃	Kalazha				0 – 50			Terrogenous deposits of inland depression, dominated by alluvial fan and braided fluvial conglomerates and sandstones in the lower part, and fluvial delta and shallow to deep lacustrine muddy deposits with coal seams in the middle and upper parts
		Qigu	152			388			
		Qiakemake				230	VI₂	VI	
	J₂	Kezilenuer	180			928			
		Yangxia				330			Widespread angular unconformity extensively uplifted along the central and southern parts of the basin
	J₁	Ahe	205	T₄₃		307	VI₁		
Triassic	T₃	Taliqike				544 – 837	V₃		Inland depression and foreland basins
		Huangshanjie	230			168 – 290	V₂	V	Alluvial, fluvial and deltaic, and lacustrine deposits, with thick lacustrine dark mudstones
	T₂	Kelamayi	240			424 – 572			
	T₁	Ehuobulake	250	T₄		37 – 292	V₁		
Permian	P₃	Shajingzi	257			600 – 700			Clatonic inland depression
	P₂	Kaipaizileike	277			430–1117	IV₂		Clastic shoreline, fluvial and deltaic deposits, with extensive volanic rock in the upper part of the Permian
		Kupukuziman				369 – 458			
	P₁	Nanzha	295	T₅₂		20 – 50			
Carboniferous	C₂	Xiaohaizhi	320			25 – 185		IV	
	C₁	Calashayi	354			65 – 270	IV₁		Clatonic inland depression clastic shoreline, fluvial and deltaic deposits Intensive uplift and erosion
		Bachu				186 – 429			
Devonian	D₃	Donghetang	372	T₅₃		15 – 192			Angular unconformity
	D₁₋₂	Keziertage	410	T₅₄		200 – 600	III₂		Clatonic inland depression; clastic shoreline, fluvial and deltaic deposits Local uplift
Silurian	S	Yimugantawu				300 – 700		III	Clatonic inland depression, clastic shoreline, deltaic and shallow marine deposits Intensive uplife and erosion
		Tataaiertage				100 – 600	III₁		
		Kepingtage	438	T₅		100 – 800			Angular unconformity
Ordovician	O₃	Tickereke		T₂₅,₇		500 – 3000	II₆		Convergent continental margin and retroarc foreland setting; Open carbonate platform, shelf and slope deposits with widely developed reef and shoal complexes of platform margin
		Sangtamu							
		Lianglitage				45 – 102	II₅		Local uplift Angular unconformity along the central basin
		Tumuxiuke				30 – 60			
	O₂	Yijianfang		T₂₅,₇		50 – 100			
		Yingshan				600 – 900	II₄		Open and restricted carbonate platform, limestone and dolomitic limestone
	O₁	Penglaiba	490	T₂₆		250 – 600		II	Parallel and minor angular unconformity
Cambrian	Є₃	Xiaqiulitage	500			291 – 1783	II₃		Divergent continental margin Restricted carbonate platform, dolomite, dolomitic limestone, gypsum and gypsiferous mudstone, with shelf mudstone and phosphorite at the base
	Є₂	Awatake				66 – 330	II₂		
		Shayilike	513	T₉₇		33 – 99			
	Є₁	Wusongger				75 – 412	II₁		
		Xiaoerbulake				40 – 240			
		Yuertushi	543	T₉₈		8 – 35			Parallel and minor angular unconformity
Sinian	Z₂					50 – 200	I₂	I	
	Z₁			T₉₁₀		0 – 3000	I₁		

Fig. 1.4 Generalized Phanerozoic tectonostratigraphy of the Tarim Block, showing the unconformity-bounded sequences and the evolution of the deposition and the dynamic setting (After Lin et al., 2012).

From the Silurian to the Lower–Middle Devonian contact with Pre-Silurian was a widespread angular unconformity caused by collision between Tarim Block and North Kunlun Block (Fig. 1.4). This tectonic event also ended the development of the Late Ordovician deepwater basin. The Silurian basin is filled with fluvial, deltaic, or clastic littoral deposits, and the Lower to Middle Devonian is composed of fluvial and littoral deposits with red beds. At the end of the Middle Devonian, the Tarim Block suffered strong deformation and denudation with the largest erosion amount of up to 3000–5000 m along the northern and northeastern margin, forming an angular unconformity distributed over most parts of the Tarim Block.

During the Late Devonian, the tectonic geomorphology was generally high in the northeast and low in the southwest from the Late Devonian to the Carboniferous. This significantly constrained the distribution of the Late Devonian to the Early Carboniferous paleogeography of the block. As transgression took place northeastwards, the Donghetang Formation was composited by shoreline and wave-dominated deltaic deposits predominated by clean quartz sandstones. During the Early Permian, one notable feature of the Tarim Block was the wide occurrence of the Early Permian intra-plate magmatism in which the magmatic rocks were made up mainly of basaltic rocks, including basalts, diabase, basaltic andesite, ultramafic rocks, etc. The residual area of the magmatic rocks can reach a size of about 2.5×10^5 m^2, and this magmatism resulted in a regional unconformity and onlap contact between the earliest Permian and Carboniferous strata with denudation of the Carboniferous and the deposition of the earliest Permian strata. During the Permian, the deposition of the Tarim Block evolved from a marine into a fluvial and lacustrine intercontinental basin, which caused the uplift and paleogeographic change at the beginning of the Permian.

The Triassic is made up of terrigenous clastic deposits, including alluvial and lacustrine deposits (Ji et al., 2003), which can be classified into three subordinate tectonostratigraphic units or sequences (Fig. 1.5). There was an inland depression in the central basin as well as the Kuqa foreland depression along the northwestern margin of the basin; the formation of the latter was related genetically to the collision of the South Tianshan orogeny and the Tarim Block during the Triassic (Zhang et al., 2007a, b). In the Kuqa depression, the Triassic deposits formed a wedge that thickens toward the northern marginal foredeep.

Jurassic sediments deposited on pre-Jurassic ones with a lack of unconformity, which is a block scale of unconformity, and heavy erosion, can be recognized and traced on seismic profiles in most parts of the block, particularly in the northeastern part of it. The nature of the Jurassic Tarim Block was in regional extension although there has long been a debate on it (Sobel, 1999; Wu et al., 2005). The Lower Jurassic Ahe Formation is composed of alluvial fan and braided fluvial conglomerates and sandstone. The overlying Yangxia Formation of the Lower Jurassic is dominated by shallow and deep muddy lacustrine deposits. There is an unconformity at the base of the Middle Jurassic, and the fluvial sandstones of the Kezhilenuer Formation of the Middle Jurassic irregularly overlying the thick mudstones of lacustrine deposits of the Yangxia Formation in most parts of the block. The Lower Cretaceous includes alluvial, fan delta or braider river delta, shoreline, and shallow or semi-deep lacustrine and arid salt lake deposits. The Upper Cretaceous in the Southwest Depression contains thin beds of fossiliferous mudstones, bioclastic limestone, reef limestone, and gypsum mudstone of marine and lagoon, tidal deposits, which are associated with shoreline clastic and deltaic deposits.

The Eogene and Neogene can be subdivided into four composite sequences in the Kuqa depression, and each of them ranging from 400 to 1000 m in thickness, are confined by unconformities and are made up of regional depositional cycles evolving from alluvial or fluvial to lacustrine and finally to fluvial deposits. The basal surface of the Kumugelimu Formation is a widespread minor angular unconformity, and the lower part of the Kumugelimu Formation has thick beds of alluvial fan deposit overlying the Lower Cretaceous. It has been suggested that the formation of these sequences is controlled by the foreland tectonic process from flexural subsidence caused by thrust loading to rebounded uplift due to the erosion and stress release (Fig. 1.2). In the central foredeeps, the upper part of the Eogene contains thick beds of gypsum mudstones and salt beds and some thin beds of limestones deposited in a lagoon or embayment environment related to instantaneous marine transgression. Up until the Miocene, the compression from orogenic belts intermittently increased and the foredeeps along the foreland of the Tianshan and Kunlun Mountains were thrusted, leading to a series of faulted and folded structural belts.

1.1.3 Igneous Activities

Major episodes of the igneous-metamorphic activities in the Tarim Block include those during the Late Archean, the Paleoproterozoic, the Late Mesoproterozoic to the Late Neoproterozoic, and the Early Permian.

The Late Archean include the tonalitic granitic rocks (as well as some gabbro enclaves in granitic rocks). In most places, the Archean rocks outcrop as stripes or lenses with variable dimensions that are tectonically enclosed within the Paleoproterozoic paragneiss; both of them generally show foliations that are concurrent to each other (Zhang et al., 2012). The Late Archean gneissic granites in the Quruqtagh area, as well as some gabbro enclaves in the granites, were termed as the Tuogelakebulake complex. Petrographically, the Late Archean rocks could be divided into three groups, i.e., the TTG (tonalite, trondhjemite, granodiorite as well as granites), calc-alkaline granites and high Ba–Sr granites.

The Early Paleoproterozoic igneous activity was documented in the southwestern, northern and eastern parts of the Tarim Block. The rock types include A–S granites with gabbro enclaves on the southwestern margin, calc-alkaline granites in the Quruqtagh area, gneiss granites and mafic dikes in the eastern Tarim Block (Zhang et al., 2007b; Lu et al., 2008; Long et al., 2010). According to geochemical studies, these igneous rocks were formed in an extensional environment (Zhang et al., 2007b; Lu et al., 2008). Lei et al. (2012) reported 1944–1934 Ma granites in the Xishankou area near Korla City, and these granites show continental-arc-type geochemical signatures.

The gneissic granites from the Late Mesoproterozoic period to the Neoproterozoic mostly outcrop in the Quruqtagh and Altyn–Dunhuang areas (Deng et al., 2008; Lu et al., 2008; Shu et al., 2011) and ca. 1.0 Ga diorite of arc signatures was reported in the central–Tarim area (Li et al., 2005). In combination with the coeval metamorphism (1.0–0.9 Ga) around the Tarim Block, it was suggested that this phase of igneous-metamorphic activity was intimately related to the assembly of the Rodinia supercontinent. Neoproterozoic igneous activities along the northern margin of the Tarim Block could be broadly divided into four phases. They are as follows: (1) the ca. 820–800 Ma ultramafic–mafic–carbonatite complex and voluminous adakitic granites and mafic dike swarm; (2) the ca. 780–760 Ma tholeiitic ultramafic–mafic complex and voluminous mafic dike swarm; (3) the ca. 740–735 Ma bimodal complex and bimodal volcanic series; and (4) the ca. 650–635 Ma mafic dikes (Chen et al., 2004; Zhan et al., 2007; Zhu et al., 2008; Zhang HA et al., 2009; Zhang CL, 2010a, 2012). The presence of voluminous mafic rocks with an arc-like signature and the calc-alkaline granites in this area suggest that there was a partial melting of the enriched lithospheric mantle and crust in response to the Rodinian breakup (Zhang et al., 2013). The formation of enriched sources may result from the interaction between the overlying lithosphere and the oceanic crust due to subduction beneath the Tarim continent during the Grenvillian orogeny (Zhang C.-L. et al., 2011, 2012, 2013).

The Early Permian magmatic rocks display a great variety in lithology, ranging from ultramafic, mafic to felsic compositions and all of them exhibit typical within-plate affinities in geochemistry.

1.2 Brief Introduction to LIPs

The term "Large Igneous Province" (LIP) was initially proposed by Coffin and Eldholm (1991, 1992, 1993a, b, 1994) to represent a variety of mafic igneous provinces with areal extents >0.1 Mkm^2 that represented "massive crustal emplacements of predominantly mafic (Mg- and Fe-rich) extrusive and intrusive rock, and originated via processes other than 'normal' seafloor spreading. As physical manifestations of mantle processes, these global phenomena include continental flood basalts, volcanic passive margins, oceanic plateaus, submarine ridges, seamount groups and ocean basin flood basalts". But, Bryan and Ernst's (2008) revised definition emphasizes four attributes: large volume, short duration or pulsed character of the igneous events, and an intraplate setting or geochemistry. The revised definition of LIPs is as follows: "Large Igneous Provinces are magmatic provinces with areal extents >0.1 Mkm^2, igneous volumes >0.1 Mkm^3 and maximum life spans of ~50 Ma that have intraplate tectonic settings or geochemical affinities, and are characterized by igneous pulse(s) of short duration (~1–5 Ma), during which a large proportion (>75%) of the total igneous volume has been emplaced." This definition emphasizes that LIPs are mainly mafic magmatic provinces having generally subordinate ultramafic components; that substantial volumes of silicic magmatism are often an integral part of continental LIPs; and that a few continental LIPs are mainly silicic (Fig. 1.5). In the new definition, seamounts, seamount groups, submarine ridges and anomalous seafloor crust are no longer considered as LIPs. Although many of these are spatially-related features post-dating an LIP event, they are constructed by long-lived melting anomalies in the mantle at lower emplacement rates, and contrast with the more transient, high magma emplacement rate characteristics of the LIP event. Many LIPs emplaced in both continental and oceanic realms are split and rifted apart by a new ridge-spreading center, which reinforces the link with mid-ocean ridges as a post-LIP event. Three new types of igneous provinces are now included in the LIP inventory to accommodate the recognition of a greater diversity of igneous compositions and preserved expressions of LIP events since the Archean: 1) a giant diabase/dolerite continental dike swarm, sill and mafic–ultramafic intrusion-dominated provinces; 2) Silicic LIPs; and 3) tholeiite–komatiite associations.

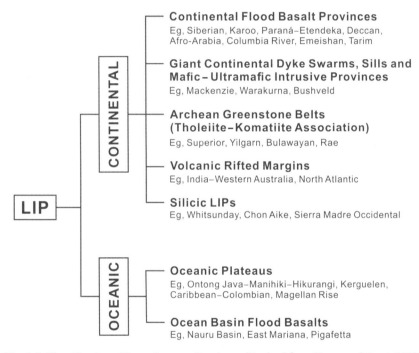

Fig. 1.5 Classification of Large Igneous Provinces (Revised from Bryan and Ernst, 2008)

LIPs can occur throughout the Earth's history (Fig. 1.6; Bryan and Ernst, 2008). Those of the Mesozoic and Cenozoic ages are those that are relatively well-preserved and best studied. They most conspicuously include both continental and oceanic flood basalts, and important examples include the 62–56 Ma North Atlantic Igneous Province (NAIP), the 122 Ma "greater" Ontong Java event in the Pacific Ocean, the 182 Ma Karoo–Ferrar event and the 250 Ma Siberian Traps. The continental flood basalts are commonly associated with volcanic rifted margins. Paleozoic and Proterozoic LIPs are typically more deeply eroded, exposing their plumbing system of giant dike swarms, sill provinces and layered intrusions, and an example of the 1270 Ma Mackenzie giant radiating dike swarm of the Canadian Shield fans over 100° of arc and extends for more than 2300 km from its focal point. Flood basalts also occur in the Archean; however, most Archean mafic–ultramafic magmatism occurs as deformed and fragmented packages termed greenstone belts. Those Archean greenstone belts that contain thick tholeiite sequences with minor komatiites are excellent candidates for LIPs. The Rae craton of northern Canada is a typical example which extends for more than 1000 km and consists of 2730–2700 Ma komatiite-bearing greenstone belts of the Woodburn Lake, Prince Albert, and Mary River groups (Fig. 1.6). LIPs have been recognized on Mars, Venus and the Moon where they provide complementary information to that from those on the Earth (Head and Coffin, 1997; Ernst and Desnoyers, 2004).

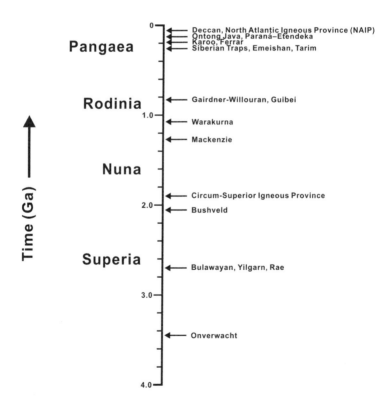

Fig. 1.6 Age spectrum of selected LIP events through time (Revised from Bryan and Ernst, 2008)

LIPs have close links with ore deposits, climatic changes and extinction events (Ernst et al., 2005). There are strong links between LIPs and Ni–Cu–PGE (platinum group element) ore deposits. For example, the Noril'sk deposits which produce 70% of the world's palladium is linked to the 250 Ma Siberian Trap event (Naldrett, 2004). Emplacement of an LIP may release massive amounts of SO_2 into the atmosphere, causing global cooling and acid rain, and CO_2, which has a strong greenhouse effect (Veevers, 1990; Campbell et al., 1992; Kerr, 1998; Wignall, 2001; Condie, 2001; Ernst and Buchan, 2003). Furthermore, a minor temperature increase can potentially trigger a massive gas hydrate melting and thus an LIP event can have an effect far greater than its direct contribution to climate change (Wignall, 2001; Jahren, 2002). Occanic LIPs can interrupt ocean circulation patterns, and cause displacement of water onto continental shelves (Kerr, 1998; Wignall, 2001). The end-Cretaceous extinction has been linked with the Chicxulub Craton, the Yucatan Peninsula, Mexico and the Deccan Trap. Other extinction events have been linked more tenuously to impacts; for example, the end-Triassic and end-Permian extinction events are linked to the Siberian Traps and the Central Atlantic Magmatic Province.

LIP clusters also have links with supercontinent breakups and juvenile crust production (Ernst et al., 2005). Spatial clusters of LIPs have been linked to supercontinental breakups (Storey, 1995; Li et al., 2003; Maruyama et al., 2007). Specifically, at least 5 LIPs have been linked to the progressive breakup of Gondwana (Storey, 1995), and several are linked to the breakup of Rodinia supercontinent (Fig. 1.8; Li et al., 2003; Maruyama et al., 2007).

1.3 Research History on the Early Permian Tarim Large Igneous Province (Tarim LIP)

One notable feature of the Tarim Block is the wide occurrence of the Early Permian intraplate magmatism (Fig. 1.7 and Fig. 1.8), and the magmatic rocks were constituted mainly of basaltic rocks including basalts, diabase, basaltic andesite, ultramafic rocks, etc. The residual distribution area of the magmatic rocks can reach about 2.5×10^5 km^2 (Fig. 1.9), and the residual thickness is from dozens of meters to several hundred meters (Yang et al., 1996, 2005, 2006a, 2007a, b; Chen et al., 1997a, 1998; Jia, 1997; Jia et al., 2004). With the scale as large as the Emeishan basalt, we regard the Tarim Early Permian intraplate magmatism as a new Large Igneous Province (Tarim LIP).

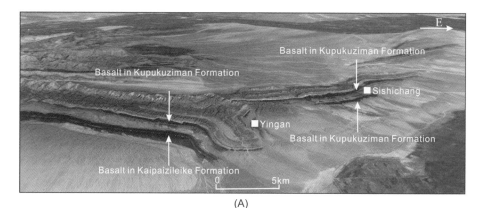

Fig. 1.7 The remote sensing images showing the distribution of basalts in the Yingan and Sishichang area (A) and basalts profile of the Kupukuziman Formation in Sishichang area (B)

Tarim Permian intraplate basalts were first reported by Wang and Liu (1991), who studied the Permian basalts in Kaipaizileike section outcropped in Keping County, northwest of the Tarim Block. Wang and Liu (1991) believed that the basalts were formed in the continental rift environment during the Late Carboniferous to Early Permian, and that they belong to subalkaline basalts at the lower part and to alkali basalt at the upper part. However, it is hard to know about their distribution and geodynamic importance because most of the basaltic rocks in the basin are buried by the upper strata. Large-scale, systematic studies of the Tarim Permian intraplate basalts were carried out by a group from the School of Earth Sciences, Zhejiang University in 1992 (Yang et al., 1994, 1996, 2005, 2006a, b, 2007a, b; Chen et al., 1997a, b, 1998; Li et al., 2008). They investigated basalts from both outcrops in the marginal area and boreholes in the interior area of the Tarim Block. The results show that the Tarim Permian magmatism was

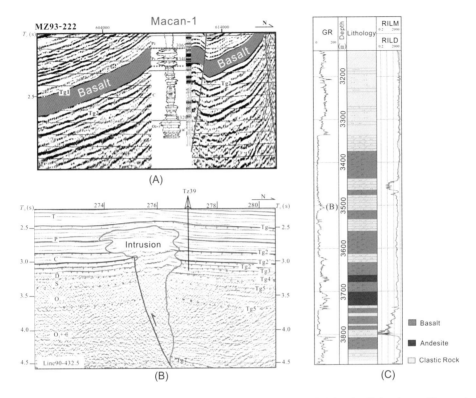

Fig. 1.8 Early Permian basalt (A) and Permian intrusion (B) in the Seismic profile, and borehole histogram of Tz-47 (C)

mainly made up of basalts with synchronous diabase, alkali granite, and ultrabasic rocks. The basalts are mainly in an alkali and subalkaline series with minor tholeiites. The results of whole rock K–Ar and Ar–Ar isotopic ages show that the magmatism occurred mainly in the Early Permian period with ages ranging from 290 to 277 Ma. The SHRIMP (sensitive high resolution ion microprobe) zircon U–Pb data of the basalts from the Keping area indicated that it first erupted at about 290 Ma and lasted until about 288 Ma (Yu et al., 2011; Li et al., 2011). From the study of the Yingan section in the Keping area, the chemical composition of basalts has a richer upward trend of K, Fe, and P. Together with the results of trace-element and rare-earth element, the group refers to the basalts as continental flood basalts (CFBs) which should have a close relation with mantle plume activities. In the meantime, the basalts, diabase, and ultrabasic rocks from outcrops in the Keping and Bachu areas were also investigated by many other authors (Jiang et al., 2004a, b, c). Jiang et al. (2004a, b, c) noted that the ultramafic

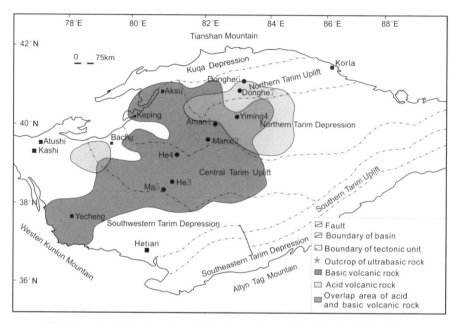

Fig. 1.9 The distribution of the Tarim Large Igneous Province (Tarim LIP)

rocks may have sourced from the D″ layer above the core-mantle boundary. Systematic studies of different units on the Tarim LIP with different perspectives have been carried out since 2006. There are representative studies of Tarim LIP that focus on the systematic studies of the geochronology and geochemistry for the Permian basalts (Li Y. et al., 2007; Li Z.-L. et al., 2008; Zhang et al., 2009; Zhou et al., 2009; Chen et al., 2010; Tian et al., 2010; Yu et al., 2010, 2011; Zhang et al., 2010a, b; Li Y.-Q. et al., 2012a; Li H.-Y. et al., 2013; Wei et al., 2014; Xu et al., 2014), intermediate-acid rocks (Chen H.-L. et al., 1998; Chen M.-M. et al., 2010; Yang et al., 2006b; Li Y. et al., 2007; Tian et al., 2010; Shangguan et al., 2011), basic and ultrabasic dikes and ultramafic rocks (Jia et al., 2001; Li Y. et al., 2007; Sun et al., 2007; Zhang C.-L. et al., 2008; Zhang H.-A. et al., 2009; Li Y.-Q. et al., 2010, 2012b; Zhou et al., 2009); the time sequence of different units in the Tarim LIP (Chen et al., 2009; Li Z.-L. et al., 2011); the comparative studies on the Tarim LIP with the Permian magmatism in adjacent areas (Yu et al., 2009; Zhang et al., 2010a, b; Qin et al., 2011; Xia et al., 2012); the relationship between the Tarim LIP and the formation of mineral resources and environmental change (Zhu et al., 2005; Chen et al., 2006; Yang et al., 2006a; Kang, 2008; Cao et al., 2013;

Li et al., 2014). After about twenty years of systematic studies, we found that the Tarim LIP has the following features:

(1) The Tarim LIP is characterized by various types of magmatic rocks and a wide area of distribution, and the area of residual distribution of the magmatic rocks is over 2.5×10^5 km^2 with that of basalts reaching about 2.0×10^5 km^2 (Yang et al., 2005, 2006a).

(2) The timing of the Tarim LIP igneous suites is between 290 and 275 Ma (Chen et al., 2009; Li Z.-L. et al., 2011), with the large scale of basaltic magma that erupted between ca. 290 and 288 Ma (Yu et al., 2011).

(3) The basalts from the Keping area have an OIB-like (OIB means oceanic island basalt) trace element pattern with enrichments in LILE (large ion lithophile elements) and HFSE (high field strength elements), a relatively high ^{87}Sr/^{86}Sr$_i$ and negative $\varepsilon_{Nd}(t)$ value suggesting that they were derived from an enriched mantle (Zhou et al., 2009; Zhang et al., 2010a), and should have a close relationship with the interaction of the mantle plume and the lithosphere (Yu et al., 2009, 2011; Zhang Y.-T. et al., 2010; Wei and Xu, 2013).

(4) The generation of the intermediate-felsic volcanic rocks in the northern Tarim Block, Bachu syenites (A-type granite) and syenitic porphyry and Piqiang A-type granite were correlated to mantle plume and the appearances of A-type granite means the end of Tarim LIP magmatism (Chen et al., 1998; Yang et al., 2006b; Sun et al., 2007; Zhang et al., 2010a).

(5) All of the geological characteristics, such as the large scale of crustal uplift (Chen et al., 2006; Li et al., 2014), the similar geochemical features of the ultrabasic rocks to the picrite (Yang et al., 2007b; Tian et al., 2010), and the large scale of dike swarms and V–Ti magnetite deposit give support to the theory that the Tarim LIP is related to the mantle plume.

(6) The Early Permian basic and ultrabasic rocks distributed in the Tarim, Junggar and Tuha Basins and in the Altay orogen may be the different parts of one tremendous LIP over 6×10^5 km^2 (Zhang et al., 2010a, b; Qin et al., 2011; Zhang C.-L. et al., 2013a, b; Zhang D.-Y. et al., 2013; Xu et al., 2014).

(7) The Tarim LIP magmatism had led to the formation of large V–Ti magnetite deposits and the change in the Permian depositional environment (Zhu et al., 2005; Yang et al., 2005; Chen et al., 2006; Kang, 2008; Cao et al., 2013).

(8) The early-stage magmatism of the Tarim LIP mainly derives from the melting of the Tarim sub-continental lithospheric mantle (SCLM) caused by the

heating of the mantle plume, and the late-stage magmatism of the Tarim LIP mainly derives from the decompression melting of the Tarim mantle plume (Yu, 2009; Xu et al., 2014).

References

Bryan, S.E., Ernst, R.E., 2008. Revised definition of Large Igneous Provinces (LIPs). Earth-Science Reviews, 86(1-4): 175-202.

Campbell, I.H., Czamanske, G.K., Fedorenko, V.A., Hill, R.I., Stepanov, V., 1992. Synchronism of the Siberian Traps and the Permian–Triassic boundary. Science, 258(5089): 1760-1763.

Cao, J., Wang, C.-Y., Xing, C.-M., Xu, Y.-G., 2013. Origin of the Early Permian Wajilitag igneous complex and associated Fe–Ti oxide mineralization in the Tarim Large Igneous Province, NW China. Journal of Asian Earth Sciences. http://dx.doi.org/10.1016/j.jseaes.2013.09.014.

Chen, H.-L., Yang, S.-F., Dong, C.-W., Jia, C.-Z., Wei, G.-Q., Wang, Z.-G., 1997a. The discovery of Permian basic rock belt in Tarim Basin and its tectonic meaning. Geochimica, 26: 77-87 (in Chinese with English abstract).

Chen, H.-L., Yang, S.-F., Dong, C.-W., Zu, G.-Q., Jia, C.-Z., Wei, G.-Q., Wang, Z.-G., 1997b. Research of geological thermal event in the Tarim Basin. Chinese Science Bulletin, 47(7): 580-584.

Chen, H.-L., Yang, S.-F., Jia, C.-Z., Dong, C.-W., Wei, G.-Q., 1998. Confirmation of Permian intermediate-acid igneous rock zone and a new understanding of tectonic evolution in the northern part of the Tarim Basin. Acta Mineral. Sin., 18(3): 370-376 (in Chinese with English abstract).

Chen, H.-L., Yang, S.-F., Wang, Q.-H., Luo, J.-C., Jia, C.-Z., Wei, G.-Q., Li, Z.-L., He, G.-Y., Hu, A.-P., 2006. Sedimentary response to the Early–Mid Permian basaltic magmatism in the Tarim plate. Geology in China, 33(3): 545-552 (in Chinese with English abstract).

Chen, H.-L., Yang, S.-F., Li, Z.-L., Yu, X., Luo, J.-C., He, G.-Y., Lin, X.-B., Wang, Q.-H., 2009. Spatial and temporal characteristics of Permian Large Igneous Province in Tarim Basin. Xinjiang Petroleum Geology, 30(2): 179-182 (in Chinese with English abstract).

Chen, M.-M., Tian, W., Zhang, Z.-L., Pan, W.-Q., Song, Y., 2010. Geochronology of the Permian basic-intermediate-acidic magma suite from Tarim, Northwest China and its geological implications. Acta Petrol. Sin., 26(2): 559-572 (in Chinese with English abstract).

Chen, Y., Xu, B., Li, Y., 2004. First mid-Neoproterozoic paleomagnetic result from the Tarim Basin (NW China) and their geodynamic implications. Precambrian Res., 133: 271-281.

Coffin, M.F., Eldholm, O., 1991. Large Igneous Provinces: JOI/USSAC workshop report. The University of Texas at Austin Institute for Geophysics Technical Report, p. 114.

Coffin, M.F., Eldholm, O., 1992. Volcanism and continental break-up: A global compilation of Large Igneous Provinces. In: Storey, B.C., Alabaster, T., Pankhurst, R.J. (Eds.), Magmatism and the Causes of Continental Break-up. Geological Society of London Special Publication, 68: 17-30.

Coffin, M.F., Eldholm, O., 1993a. Scratching the surface: Estimating dimensions of large igneous provinces. Geology, 21(6): 515-518.

Coffin, M.F., Eldholm, O., 1993b. Large Igneous Provinces. Scientific American, 269: 42-49.

Coffin, M.F., Eldholm, O., 1994. Large Igneous Provinces: Crustal structure, dimensions, and external consequences. Reviews of Geophysics, 32(1): 1-36.

Condie, K.C., 2001. Mantle Plumes and their Record in Earth History. Cambridge University Press, p. 320.

Deng, X.-L., Shu, L.-S., Zhu, W.-B., 2008. Geochronology of the Precambrian structure magmatism deformation along the Xindi fault in Xinjiang, NW China. Acta Petrologica Sinica, 24: 2800-2808 (in Chinese).

Ernst, R.E., Buchan, K.L., 2003. Recognizing mantle plumes in the geological record. Annual Review of Earth and Planetary Sciences, 16(16): 469-523.

Ernst, R.E., Desnoyers, D.W., 2004. Lessons from Venus for understanding mantle plumes on Earth. Physics of the Earth and Planetary Interiors, 146(1-2): 195-229.

Ernst, R.E., Buchan, K.L., Campbell, I.H., 2005. Frontiers in Large Igneous Province research. Lithos, 79(3-4): 271-297.

Head, J.W., Coffin, M.F., 1997. Large Igneous Provinces: A planetary perspective. In: Mahoney, J.J., Coffin, M.F. (Eds.), Large Igneous Provinces: Continental, Oceanic, and Planetary Flood Volcanism. AGU Geophys. Monogr. Ser., 100: 411-438.

Hu, A.-Q., Jahn, B.M., Zhang, G.-X., Chen, Y.-B., Zhang, Q.-F., 2000. Crustal evolution and Phanerozoic crustal growth in northern Xinjiang: Nd isotopic evidence. Part I. Isotopic characterization of basement rocks. Tectonophysics, 328(1-2): 15-51.

Jahren, A.H., 2002. The biogeochemical consequences of the mid-Cretaceous superplume. In: Condie, K.C., Abbot, D., Des Marais, D.J. (Eds.), Superplume Events in Earth's History: Causes and Effects. J. Geodyn., 34(Special Issue): 177-191.

Ji, Y.-L., Ding, X.-S., Li, X.-C., 2003. Triassic paleogeographyand sedimentary facies of the Kuqa depression, Tarim Basin. Journal of Geomechanics, 19(13): 268-274 (in Chinese with English abstract).

Jia, C.-Z., 1997. Tectonic Characteristics and Oil-Gas, Tarim Basin. Beijing: Petroleum Industry Press (in Chinese).

Jia, C.-Z., Wei, G.-Q., 2002. The tectonic characteristics of the Tarim Basin and its oil-gas potential. Chinese Science Bulletin, 47: 1-8 (in Chinese).

Jia, C.-Z., Yang, S.-F., Chen, H.-L., Wei, G.-Q., 2001. The Evolution of Tethyan Tectonic Domain and Natural Gas Accumulation in the Basin Groups at the Northern Side of Tethyan. Beijing: Petroleum Industry Press (in Chinese).

Jia, C.-Z., Zhang, S.-B., Wu, S.-Z., 2004. Stratigraphy of the Tarim Basin and Adjacent Areas. Beijing: Science Press, pp. 190-289 (in Chinese).

Jiang, C.-Y., Zhang, P.-B., Lu, D.-R., Bai, K.-Y., 2004a. Petrogenesis and magma source of the ultramafic rocks at Wajilitag region, western Tarim Plate in Xinjiang. Acta Petrol. Sin., 20(6): 1433-1444 (in Chinese with English abstract).

Jiang, C.-Y., Jia, C.-Z., Li, L.-C., Zhang, P.-B., Lu, D.-R., Bai, K.-Y., 2004b. Source of the Fe-riched-type high-Mg magma in Mazhartag region, Xinjiang. Acta Geol. Sin., 78(6): 770-780 (in Chinese with English abstract).

Jiang, C.-Y., Zhang, P.-B., Lu, D.-R., Bai, K.-Y., Wang, Y.-P., Tang, S.-H., Wang, J.-H., Yang, C., 2004c. Petrology, geochemistry and petrogenesis of the Kalpin basalts and their Nd, Sr and Pb isotopic compositions. Geol. Rev., 50(5): 492-500 (in Chinese with English abstract).

Kang, Y.-Z., 2008. Characteristic of the Carboniferous–Permian volcanic rocks and hydrocarbon accumulations in two great basins, Xinjiang area. Petrol. Geol. Exper., 30: 321-327 (in Chinese).

Kerr, A.C., 1998. Oceanic plateau formation: A cause of mass extinction and black shale deposition around the Cenomanian–Turonian boundary. Journal of the Geological Society of London, 155: 619-626.

Lei, R.-X., Wu, C.-Z., Chi, G.-X., Chen, G., Gu, L.-X., Jiang, Y.-H., 2012. Petrogenesis of the Palaeoproterozoic Xishankou pluton, northern Tarim Block, Northwest China: Implications for assembly of the supercontinent Columbia. International Geology Review, 54(15): 1829-1842.

Li, D.-X., Yang, S.-F., Chen, H.-L., Cheng, X.-G., Li, K., Jin, X.-L., Li, Z.-L., Li, Y.-Q., Zou, S.-Y., 2014. Late Carboniferous crustal uplift of the Tarim plate and its constraints on the evolution of the Early Permian Tarim Large Igneous Province. Lithos, 204: 36-46.

Li, H.-Y., Huang, X.-L., Li, W.-X., Cao, J., He, P.-L., Xu, Y.-G., 2013. Age and geochemistry of the Early Permian basalts from Qimugan in the southwestern Tarim Basin. Acta Petrologica Sinica, 29(10): 3353-3368 (in Chinese with English abstract).

Li, Y., Su, W., Kong, P., Qian, Y.-X., Zhang, K.-Y., Zhang, M.-L., Chen, Y., Cai, X.-Y., You, D.-H., 2007. Zircon U–Pb ages of the Early Permian magmatic rocks in the Tazhong–Bachu region, Tarim basin by LA-ICP-MS. Acta Petrologica Sinica, 23(5): 1097-1107 (in Chinese with English abstract).

Li, Y.-J., Song, W.-J., Wu, G.-Y., Wang, Y.-F., Li, Y.-P., Zheng, D.-M., 2005. Jinning granodiorite and diorite deeply concealed in the central Tarim Basin. Science China, Series D, 48: 2061-2068 (in Chinese).

Li, Y.-Q., Li, Z.-L., Sun, Y.-L., Chen, H.-L., Yang, S.-F., Yu, X., 2010. PGE and geochemistry of Wajilitag ultramafic cryptoexplosive brecciated rocks from Tarim Basin: Implications for petrogenesis. Acta Petrologica Sinica, 26(11): 3307-3318 (in Chinese with English abstract).

Li, Y.-Q., Li, Z.-L., Sun, Y.-L., Santosh, M., Langmuir, C.H., Chen, H.-L., Yang, S.-F., Chen, Z.-X., Yu, X., 2012a. Platinum-group elements and geochemical characteristics of the Permian continental flood basalts in the Tarim Basin, Northwest China: Implications for the evolution of the Tarim Large Igneous Province. Chemical Geology, 328: 278-289.

Li, Y.-Q., Li, Z.-L., Chen, H.-L., Yang, S.-F., Yu, X., 2012b. Mineral characteristics and metallogenesis of the Wajilitag layered mafic–ultramafic intrusion and associated Fe–Ti–V oxide deposit in the Tarim Large Igneous Province, Northwest China. Journal of Asian Earth Sciences, 49: 161-1749.

Li, Z.-L., Yang, S.-F., Chen, H.-L., Langmuir, C.-H., Yu, X., Lin, X.-B., Li, Y.-Q., 2008. Chronology and geochemistry of Taxinan basalts from the Tarim Basin: Evidence for Permian plume magmatism. Acta Petrologica Sinica, 24(5): 959-970 (in Chinese with English abstract).

Li, Z.-L., Chen, H.-L., Song, B., Li, Y.-Q., Yang, S.-F., Yu, X., 2011. Temporal evolution of the Permian Large Igneous Province in Tarim Basin in northwestern China. Journal of Asian Earth Sciences, 42(5): 917-927.

Li, Z.-X., Li, X.-H., Kinny, P.D., Wang, J., Zhang, S., Zhou, H., 2003. Geochronology of neoproterozoic syn-rift magmatism in the Yangtze Craton, South China and correlations with other continents: Evidence for a mantle superplume that broke up Rodinia. Precambrian Res., 122(1-4): 85-109.

Lin, C.-S., Li, H., Liu, J.-Y., 2012. Major unconformities, tectonostratigraphic framework, and evolution of the superimposed Tarim Basin, Northwest China. Journal of Earth Science, 23(4): 395-407.

Liu, H., Somerville, I.D., Lin, C.-S., Zhou, S.-J., 2016. Distribution of Palaeozoic tectonic superimposed unconformities in the Tarim Basin, NW China: Significance for the evolution of palaeogeomorphology and sedimentary response. Geological Journal, 51: 627-651.

Long, X.-P., Yuan, C., Sun, M., Kröner, A., Zhao, G.-C., Wilde, S., Hu, A.-Q., 2011. Reworking of the Tarim Craton by underplating of mantle plume-derived magmas: Evidence from Neoproterozoic granitoids in the Kuluketage area, NW China. Precambrian Res., 187(1-2): 1-14.

Long, X.-P., Yuan, C., Sun, M., Zhao, G.-C., Xiao, W.-J., Wang, Y.-J., Yang, Y.-H., Hu, A.-Q., 2010. Archean crustal evolution of the northern Tarim Craton, NW China: Zircon U–Pb and Hf isotopic constraints. Precambrian Res., 180(3-4): 272-284.

Lu, S.-N., 1992. The Proterozoic tectonic evolution of Kuruketage, Xinjiang. Journal of Tianjin Geology and Mineral Resources, 26-27: 279-292 (in Chinese with English abstract).

Lu, S.-N., Yuan, G.-B., 2003. Geochronology of Early Precambrian magmatic activities in Aketasdhtage, East Altyn Tagh. Acta Geological Sinica, 77: 61-68 (in Chinesewith English abstract).

Lu, S.-N., Li, H.-K., Zhang, C.-L., Niu, G.-H., 2008. Geological and geochronological evidence for the Precambrian evolution of the Tarim Craton and surrounding continental fragments. Precambrian Res., 160(1-2): 94-107.

Maruyama, S., Santosh, M., Zhao, D., 2007. Superplume, supercontinent, and post-perovskite: Mantle dynamics and anti-plate tectonics on the core-mantle boundary. Gondwana Res.,

11(1-2): 7-37.

Naldrett, A.J., 2004. Magmatic Sulfide Deposits: Geology, Geochemistry and Exploration. Berlin: Springer.

Qin, K.-Z., Su, B.-X., Sakyi, P.A., Tang, D.-M., Li, X.-H., Sun, H., Xiao, Q.-H., Liu, P.-P., 2011. SIMS zircon U–Pb geochronology and Sr–Nd isotopes of Ni–Cu-bearing mafic–ultramafic intrusions in Eastern Tianshan and Beishan in correlation with flood basalts in Tarim Basin (NW China): Constraints on a ca. 280 Ma mantle plume. American Journal of Science, 311(3): 237-260.

Shangguan, S.-M., Tian, W., Li, X.-H., Guan, P., Pan, M., Chen, M.-M., Pan, W.-Q., 2011. SIMS zircon U–Pb age of a rhyolite layer from the Halahatang area, northern Tarim, NW China: Constraint on the eruption age of major pulse of Tarim flood basalt. Acta Scientiarum Naturalium Universitatis Pekinensis, 47(3): 561-564 (in Chinese with English abstract).

Shu, L.-S., Deng, X.-L., Zhu, W.-B., Ma, D.-S., Xiao, W.-J., 2011. Precambrian tectonic evolution of the Tarim Block, NW China: New geochronological insights from the Quruqtagh domain. Journal of Asian Earth Sciences, 42(5): 774-790.

Sobel, E.R., 1999. Basin analysis of the Jurassic–Lower Cretaceous southwest Tarim Basin, Northwest China. Geological Society of America Bulletin, 111(5): 709-724.

Storey, B.C., 1995. The role of mantle plumes in continental breakup: Case histories from Gondwana land. Nature, 377: 301-308.

Sun, L.-H., Wang, Y.-J., Fan, W.-M., Peng, T.-P., 2007. Petrogenesis and tectonic significances of the diabase dikes in the Bachu area, Xinjiang. Acta Petrologica Sinica, 23(6): 1369-1380.

Tian, W., Campbell, I.H., Allen, C.M., Guan, P., Pan, W.-Q., Chen, M.-M., Yu, H.-J., Zhu, W.-P., 2010. The Tarim picrite-basalt-rhyolite suite, a Permian flood basalt from Northwest China with contrasting rhyolites produced by fractional crystallization and anatexis. Contributions to Mineralogy and Petrology, 160(3): 407-425.

Veevers, J.J., 1990. Tectonic-climatic supercycle in the billion-year plate-tectoniceon: Permian Pangean icehouse alternates with cretaceous dispersed-continents greenhouse. Sediment. Geol., 68: 1-16.

Wang, T.-Y, Liu, J.-K., 1991. A preliminary investigation on formative phase and rifting of Tarim Basin. In: Jia, R.-X. (Ed.), Research of Petroleum Geology of Northern Tarim Basin. Beijing: China University of Geoscience Press, pp. 115-124 (in Chinese).

Wei, X., Xu, Y.-G., 2013. Petrogenesis of the mafic dikes from Bachu and implications for the magma evolution of the Tarim Large Igneous Province, NW China. Acta Petrologica Sinica, 29: 3323-3335 (in Chinese with English abstract).

Wei, X., Xu, Y.-G., Feng, Y.-X., Zhao, J.-X., 2014. Plume-lithosphere interaction in the generation of the Tarim Large Igneous Province, NW China: Geochronological and geochemical constraints. American Journal of Science, 314: 314-356.

Wignall, P.B., 2001. Large igneous provinces and mass extinctions. Earth Science Reviews,

53(1): 1-33.

Wu, C.-D., Lin, C.-S., Shen, Y.-P., 2005. Composition of sandstone and heavy minerals implies the provenance of Kuqa depression in Jurassic, Tarim Basin, China. Progress in Natural Science, 15(7): 633-640.

Xia, L.-Q., Xu, X.-Y., Li, X.-M., Ma, Z.-P., Xia, Z.-C., 2012. Reassessment of petrogenesis of Carboniferous–Early Permian rift-related volcanic rocks in the Chinese Tianshan and its neighboring areas. Geoscience Frontiers, 3(4): 445-471.

Xinjiang BGUR (Xinjiang Bureau of Geology and Mineral Resources), 1993. Regional geology of the Xinjiang Uygur Autonomous Region. Beijing: Geol. Publ. House, pp. 1-48 (in Chinese).

Xu, Y.-G., Wei, X., Luo, Z.-Y., Liu, H.-Q., Cao, J., 2014. The Early Permian Tarim Large Igneous Province: Main characteristics and a plume incubation model. Lithos, 204: 20-35.

Yang, S.-F., Chen, H.-L., Dong, C.-W., Jia, C.-Z., Wang, Z.-G., 1994. The research on the Paleozoic volcanism and geothermal event. Report of the National Key Project of Science and Technology, pp. 85-101 (in Chinese).

Yang, S.-F., Chen, H.-L., Dong, C.-W., Jia, C.-Z., Wang, Z.-G., 1996. The discovery of Permian syenite inside Tarim Basin and its geodynamic significance. Geochimica, 25(2): 121-128 (in Chinese with English abstract).

Yang, S.-F., Chen, H.-L., Ji, D.-W., Li, Z.-L., Dong, C.-W., Jia, C.-Z., Wei, G.-Q., 2005. Geological process of Early to Middle Permian magmatism in Tarim Basin and its geodynamic significance. Geol. J. China Univ., 11(4): 504-511 (in Chinese with English abstract).

Yang, S.-F., Li, Z.-L., Chen, H.-L., Chen, W., Yu, X., 2006a. ^{40}Ar-^{39}Ar dating of basalts from Tarim Basin, NW China and its implication to a Permian thermal tectonic event. J. Zhejiang Univ.-SCI (A), 7(Suppl. II): 320-324.

Yang, S.-F., Li, Z.-L., Chen, H.-L., Xiao, W.-J., Yu, X., Lin, X.-B., Shi, X.-G., 2006b. Discovery of a Permian quartz syenitic porphyritic dike from the Tarim Basin and its tectonic implications. Acta Petrologica Sinica, 22(5): 1405-1412 (in Chinese with English abstract).

Yang, S.-F., Li, Z.-L., Chen, H.-L., Santosh, M., Dong, C.-W., Yu, X., 2007a. Permian bimodal dike of Tarim Basin, NW China: Geochemical characteristics and tectonic implications. Gondwana Res., 12(1-2): 113-120.

Yang, S.-F., Yu, X., Chen, H.-L., Li, Z.-L., Wang, Q.-H., Luo, J.-C., 2007b. Geochemical characteristics and petrogenesis of Permian Xiaohaizi ultrabasic dike in Bachu area, Tarim Basin. Acta Petrologica Sinica, 23(5): 1087-1096 (in Chinese with English abstract).

Yu, X., 2009. Magma Evolution and Deep Geological Processes of Early Permian Tarim Large Igneous Province. Ph.D. Thesis, Zhejiang University (in Chinese with English abstract).

Yu, X., Chen, H.-L., Yang, S.-F., Li, Z.-L., Wang, Q.-H., Lin, X.-B., Xu, Y., Luo, J.-C., 2009. Geochemical features of Permian basalts in Tarim Basin and compared with Emeishan LIPs. Acta Petrologica Sinica, 25(6): 1492-1498 (in Chinese with English abstract).

Yu, X., Chen, H.-L., Yang, S.-F., Li, Z.-L., Wang, Q.-H., Li, Z.-H., 2010. Distribution characters of Permian basalts and their geological significance in the Kalpin area, Xinjiang. J. Stratigr., 34(2): 127-134 (in Chinese with English abstract).

Yu, X., Yang, S.-F., Chen, H.-L., Chen, Z.-Q., Li, Z.-L., Batt, G.E., Li, Y.-Q., 2011. Permian flood basalts from the Tarim Basin, Northwest China: SHRIMP zircon U–Pb dating and geochemical characteristics. Gondwana Res., 20(2-3): 485-497.

Zhan, S., Chen, Y., Xu, B., Wang, B., Faure, M., 2007. Late Neoproterozoic paleomagnetic results from the Sugetbrak Formation of the Aksu area, Tarim Basin (NW China) and their implications to paleogeographic reconstructions and the snowball Earth hypothesis. Precambrian Res., 154: 143-158.

Zhang, C.-L., Dong, Y.-G., Zhao, Y., 2003. Geochemistry of Meso-Proterozoic volcanites in western Kunlun: Evidence for the plate tectonic evolution. Acta Geologica Sinica, 77: 237-245.

Zhang, C.-L., Li, X.-H., Li, Z.-X., Lu, S.-N., Ye, H.-M., Li, H.-M., 2007a. Neoproterozoic ultramafic–mafic-carbonatite complex and granitoids in Quruqtagh of northeastern Tarim Block, Western China: Geochronology, geochemistry and tectonic implications. Precambrian Res., 152: 149-169.

Zhang, C.-L., Li, Z.-X., Li, X.-H., Ye, H.-M., 2007b. Early Palaeoproterozoic high-K intrusivecomplex in southwestern Tarim Block, NW China: Age, geochemistry and implications for the Paleoproterozoic tectonic evolution of Tarim. Gondwana Res., 12: 101-112.

Zhang, C.-L., Li, X.-H., Li, Z.-X., Ye, H.-M., Li, C.-N., 2008. A Permian layered intrusive complex in the western Tarim Block, northwestern China: Product of a ca. 275-Ma mantle plume. J. Geol., 116(3): 269-287.

Zhang, C.-L., Xu, Y.-G., Li, Z.-X., Wang, H.-Y., Ye, H.-M., 2010a. Diverse Permian magmatism in the Tarim Block, NW China: Genetically linked to the Permian Tarim mantle plume. Lithos, 119(3-4): 537-552.

Zhang, C.-L., Li, Z.-X., Li, X.-H., Xu, Y.-G., Zhou, G., Ye, H.-M., 2010b. A Permian large igneous province in Tarim and Central Asian orogenic belt, NW China: Results of a ca. 275 Ma mantle plume. Geol. Soc. Am. Bull., 122 (11-12): 2020-2040.

Zhang, C.-L., Yang, D.-S., Wang, H.-Y., Takahashi, Y., Ye, H.-M., 2011. Neoproterozoic ultramfic–mafic layered intrusion in Quruqtagh of northeastern Tarim Block, NW China: Two phases of mafic igneous activity of different mantle sources. Gondwana Res., 19: 177-190.

Zhang, C.-L., Li, H.-K., Santosh, M., Li, Z.-X., Zou, H.-B., Wang, H.-Y., Ye, H.-M., 2012. Precambrian evolution and cratonization of the Tarim Block, NW China: Petrology, geochemistry, Nd-isotopes and U–Pb zircon geochronology from Archaean gabbro-TTG-potassic granite suite and Paleoproterozoic metamorphic belt. Journal of Asian Earth Sciences, 47: 5-20.

Zhang, C.-L., Zou, H.-B., Li, H.-K., 2013. Tectonic framework and evolution of the Tarim Block, NW China. Gondwana Res., doi.org/10.1016/j.gr.2012.05.009.

Zhang, D.-Y., Zhang, Z.-C., Santosh, M., Cheng, Z.-G., He, H., Kang, J.-L., 2013. Perovskite and baddeleyite from kimberlitic intrusions in the Tarim Large Igneous Province signal onset of an end-Carboniferous mantle plume. Earth Planet. Sci. Lett., 361: 238-248.

Zhang, H.-A., Li, Y.-J., Wu, G.-Y., 2009. Isotopic geochronology of Permian igneous rocks in the Tarim Basin. Chinese Journal of Geology, 44: 137-158 (in Chinese with English abstract).

Zhang, Z.-Y., Zhu, W.-B., Shu, L.-S., Su, J.-B., Zheng, B.-H., 2009. Neoproterozoic ages of the Kuluketage diabase dike swarm in Tarim, NW China, and its relationship to the breakup of Rodinia. Geological Magazine, 146: 150-154.

Zhang, Y.-T., Liu, J.-Q., Guo, Z.-F., 2010. Permian basaltic rocks in the Tarim Basin, NW-China: Implications for plume-lithosphere interaction. Gondwana Res., 18: 596-610.

Zhao, G.-C., Cawood, P.A., 2012. Precambrian geology of China. Precambrian Res., 222-223: 13-45.

Zhou, M.-F., Zhao, J.-H., Jiang, C.-Y., 2009. OIB-like, heterogeneous mantle sources of Permian basaltic magmatism in the western Tarim Basin, NW China: Implications for a possible Permian Large Igneous Province. Lithos, 113: 583-594

Zhu, W.-B., Zhang, Z.-Z., Shu, L.-S., Lu, H.-F., Sun, J.-B., Yang, W., 2008. SHRIMP U–Pb zircon geochronology of Neoproterozoic Korla mafic dikes in the northern Tarim Block, NW China: Implications for the long-lasting breakup process of Rodinia. Journal of the Geological Society, 165: 887-890.

Zhu, Y.-X., Jin, Z.-J., Lin, C.-S., Lv, X.-X., Xie, Q.-L., 2005. Relations between the early Permian magmatic rocks and hydrocarbon accumulation in the central Tarim. Petroleum Geology and Experiment, 27(1): 50-54.

2

Tempo-Spatial Features of the Tarim LIP

Abstract: The study of the tempo-spatial features of the Tarim Large Igneous Province (Tarim LIP) can provide some good evidence of the limits of one of the world's Large Igneous Provinces (LIPs). The results of some recent research on the tempo-spatial features of the Tarim LIP will be addressed in this study. Through the systematical and detailed study of the spatial distribution and the precise dating and contact relationships between the main rock units in the Tarim LIP and the host rocks, the distribution of the Tarim LIP reaches over 2.5×10^5 km^2, which is a Permian LIP in NW China. The temporal order of the Tarim LIP can be considered continental flood basaltic lavas in the Kupukuziman and Kaipaizileike formations (292–285 Ma) in the early stage, and layered mafic ultramafic complex, very rich magnesium ultramafic dike, mica-olivine pyroxenite breccia pipe, diabase dike, quartz syenite and bimodal dike (284–274 Ma) in the late stage.

Keywords: Temporal order; Spatial distribution; Field contact relationship; Chronology; Tarim Large Igneous Province (Tarim LIP)

The studies on temporal order and spatial distribution as well as on the tectonic implication for LIPs in the world provide a good contribution to geological history, geodynamics and large-scale V–Ti magnetite and Cu–Ni sulfide deposits, genetic links with the mantle plume and mass extinction, and even a good contribution to the understanding of the tectono-magmatism of the Eurasia plate and the mantle-crustal interaction in the Late Paleozoic (Coffin and Eldholm, 1994; Chung and Jahn, 1995; Wignall, 2001; Zhou et al., 2002; Morgan et al., 2004; Dobretsov, 2005; Xu et al., 2001, 2007, 2014; Halls et al., 2008; Pirajno et al., 2008; Bryan and Ernst, 2008). In this chapter, we will address the temporal framework and spatial distribution of the Tarim LIP, its stratigraphic succession, radiometric isotopic chronology and paleontology, contact relationships among huge-volume basaltic lavas, intrusive rocks such as of syenites, diabase dike swarm, tuff and volcanic breccia and sedimentary rocks; moreover, this study provides some appropriate constraints for magmatic evolution and temporal sequence.

The Tarim Basin in northwestern China, surrounded by the Tianshan, Kunlun and Altyn-Tagh orogenic belts (see Fig. 1.1), mainly consists of a Precambrian crystalline basement and Phanerozoic strata from the Ordovician to Neogene periods (Xinjiang BGMR, 1993; Jia, 1997; Zhang et al., 2003). Some important tectonothermal activities from the Archean period to the Paleozoic period have been identified in this area (Hu et al., 2000; Guo et al., 2005; Xu et al., 2005, 2009; Yang et al., 2006a, 2007; Long et al., 2010, 2011; Shu et al., 2011; Zhu et al., 2011), in which the early Permian magmatic event (known as the Tarim LIP) was regarded as the most important one (Yang et al., 2006a, b, c, 2007; Zhang C.-L. et al., 2008, 2010; Zhang Y.-T., 2010; Li et al., 2011). The Tarim LIP is widely distributed in the western and central part of the Tarim Basin, including the Keping, Bachu, Tabei (northern Tarim Basin), the Tazhong (central Tarim Basin) and the Taxinan (southwestern Tarim Basin) areas (Fig. 2.1). Large-scale continental flood basalt lavas erupted during the Early Permian period (Yang et al., 2006a, b, c; Chen et al., 2006; Li et al., 2008, 2011; Tian et al., 2010; Yu et al., 2011), constituting the main part of the Tarim LIP. A diverse assemblage of coeval intrusive rocks, such as layered mafic–ultramafic intrusions, mica-olivine pyroxenite breccia pipes, diabase and ultramafic dikes, quartz syenites, diorite, quartz syenite porphyry and bimodal dikes, occurred in the Tarim Basin.

2.1 Lithological Characteristics of Different Rock Units in the Tarim LIP

The lithological characteristics of different magmatic rock units in the Tarim LIP, including the rock units of basalts, diabases, layered intrusive rocks, breccia pipe mica-olivine pyroxenite, olivine pyroxenite, gabbro, ultramafic dike, quartz syenite, quartz syenitic porphyry, diorite and bimodal dike (Yang et al., 1996; Chen et al., 1997; Jiang et al., 2004a, b, 2006; Yang et al., 2006a, b, 2007; Zhang C.-L. et al., 2008; Li et al., 2011; Zhang D.-Y. et al., 2012, 2013; Huang et al., 2012; Xu et al., 2014; Zou et al., 2015) have been described.

2.1.1 Basalts

Widespread Permian basalts have been studied from both outcrops and boreholes in the Tarim Basin recently (Chen et al., 1997; Jiang et al., 2004b, 2006; Yang et al., 2005, 2006a; Li Z.-L. et al., 2008, 2011, 2012; Zhou et al., 2009; Tian et al., 2010; Zhang Y.-T. et al., 2010; Yu et al., 2011; Zhang P.-Y. et al., 2012; Li Y.-Q. et al., 2012, 2014; Xu et al., 2014). The outcrop sections are mainly exposed in the Keping area (e.g., the Yingan, Sishichang and Kaipaizileike sections; Fig. 2.1) and can be subdivided into the Early Permian lower Kupukuziman and upper Kaipaizileike formations (Jiang et al., 2006; Yu et al., 2011; Li et al., 2011). However, most of basaltic lavas in the basin are buried in the subsurface by the desert and/or have eroded. The borehole data indicates that the basaltic rocks range in thickness from ca. 200 to 600 m (mean ~300 m) with minor sedimentary interlayers, suggesting a volume of 7.5×10^4 km^3 (Zhou et al., 2009; Tian et al., 2010; Yu et al., 2011; Li et al., 2011). A 3D seismic cross-section suggests that the seismological thickness of the basaltic sequence interbedded with scarce sedimentary rocks in the borehole of the northern Tarim Uplift (NTU; YM8 and YM5 in Fig. 2.1) is well in excess of 2500 m (Tian et al., 2010).

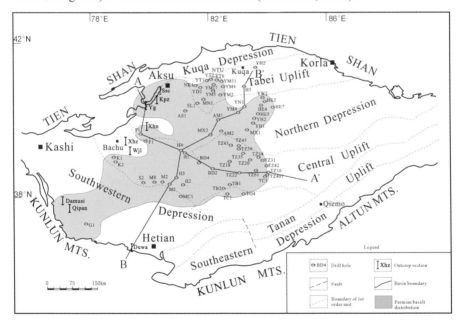

Fig. 2.1 Spatial distribution of the Permian basalts of the Tarim LIP, locations of outcrop and borehole sections of the Tarim LIP, and spatial section lines of A-A' and B-B' almost cover the whole area of the Tarim LIP in the Tarim Basin (based on field observations, boreholes and seismic data) (Simply modified from Fig. 1b of Li et al., 2011)

Note: Yg = Yingan section, Kpz = Kaipaizileike section, Ssc = Sishichang section, Xhn = Xiahenan section, Xhz = Xiaohaizi section, NTU = northern Tarim Uplift, Wjl = Wajilitag section

The Yingan section (Fig. 2.2E), located in Keping County, is one of the best exposed and preserved Permian basalt outcrops, and the basaltic layer has a field occurrence of $180°\angle 44°$ in the surrounding sedimental stratum. The basalt layers have a total thickness of ca. 530 m in this section except for small amounts of intercalated Permian sedimental layers composed of mudstones, siltstones, and mud-bearing limestones. A total of 8 basaltic lava flow units were recorded with 2 in the lower Kupukuziman Formation (Ku) and the other 6 in the upper Kaipaizileike Formation (Kai) (see Fig. 1.7). Individual basaltic units range from 10 to 70 m in thickness, with terrestrial sediments such as alternating reddish mudstone and siltstone, volcanic breccia, tuff and non-marine limestone intercalations. Several basaltic units in the Kaipaizileike Formation have well-developed columnar jointings.

Basalts are well surveyed, and lots of samples were recently collected for petrological, geochemical, Sr–Nd–Pb–Hf–PGE (platinum group elements) and zircon chronological dating study as cited in the above references. In general, the basalts are mainly dark brown, and composed of phenocrysts (10–50 modal%) of plagioclase and clinopyroxene, and there is a groundmass of fine-grained plagioclase, clinopyroxene and magnetite with minor olivine. They show holocrystalline and intersertal textures. Some basalts displayed a vesicular and amygdaloidal structure with the amygdales mainly made up of quartz and calcite. Besides, the basalts in the middle and upper part of the Kai5 unit have abundant plagioclase (ca. 40 modal%) and form clusters (Yu et al., 2011; Li Y.-Q. et al., 2012).

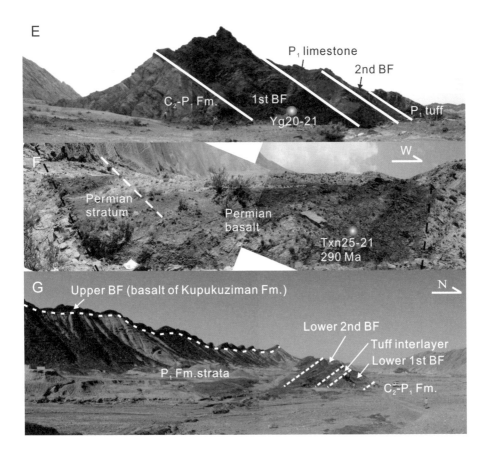

Fig. 2.2 Field contact relationships among different rock units in the Tarim LIP and country rocks. (A) Permian bimodal dikes consisting of quartz syenitic porphyry and diabase intruded into Silurian and Devonian strata in the Xiaohaizi area of Bachu County; (B) Very rich magnesium ultramafic dike and diabase intruded into the Silurian and Devonian strata in the Xiaohaizi area of Bachu County; (C) Two-episodes crossed diabase dikes intruded into the Silurian and Devonian strata in the Wajilitag area of Bachu County; (D) Diabase dike intruded into ultramafic breccia pipe in the Wajilitag area of Bachu County; (E) The relationships among basalt layers, tuff and the sedimentary strata of the upper Carboniferous and lower Permian periods in the Yingan section of the Keping area; (F) Basalt layers interlayered by lower Permian strata in the Qipan section of the Taxinan area; (G) The relationships among basalt layers, tuff and the sedimentary strata of the upper Carboniferous (marine sediment) and lower Permian (continental sediment, such as siltstone and mudstone) in the Sishichang section of the Keping area (Modified from Yang et al., 2007; Li et al., 2008., 2011)

A detailed petrographical description was made, and mineral compositions of basalts from each BU (basaltic unit) can be seen in Table 1 of Yu et al. (2011).

BU1 has a primary vesicular character, with abundant amygdaloidal infilling. Vesicles are generally elongated and exhibit consistent orientations, indicating a lava flow direction. BUs 2, 3 and 8 exhibit similar mineral compositions in both phenocrysts and groundmass, although BU3 contains a higher abundance of clinopyroxene phenocrysts. BU4 lacks clinopyroxene phenocrysts, but is dominated by an abundance of plagioclase phenocrysts. BUs 5 and 6 exhibit similar phenocryst and groundmass compositions. BU7 presents a more complex mineralogical assemblage, with the earlier flows being notable for both the presence of modal olivine and the absence of plagioclase phenocrysts, and the later flows having a high abundance of large plagioclase phenocrysts. With the exception of some aphyric basalts (notably in BU1), ophitic and diabasic textures are very common in the porphyritic basalts. Columnar jointing is observed in the basal part of most basalt flows, particularly in the Kaipaizileike section, indicating that these flows cooled quickly during the early stages of each eruptive episode. Some eruptions may also have generated pyroclastic rocks (volcanic breccia and tuff), such as those noted overlying BU2 (Yu et al., 2011).

Furthermore, Li et al. (2008) reported the basalts exposed in the Early Permian Qipan Formation of the southwestern Tarim Basin (Damusi section, see Fig. 2.2F), which show a close affinity with the Keping basalts in terms of eruption time and geochemistry. Under the microscope, basalts have a lithological change from the lower to upper parts, which shows that basalts in the lower part are fine-grained, and composed of plagioclase phenocryst and groundmass, while the latter has plagioclase microcrystal and cryptocrystalline, and they have a cryptocrystalline texture, a vesicle and some amygdaloid; basalts in the middle part become more coarse-grained, but those in the upper part become fine-grained with less vesicle. Furthermore, another basaltic layer in the Qipan Formation close to the main field section of ca. 0.5 km, has an unclear field relationship with the main section, and has a thickness of 6–8 m, outcropped in the core part of anticline of the Permian strata with the south side of the anticline being well outcropped and the north side covered.

In the northern Tarim uplift (NTU), Tian et al. (2010) discovered that most basaltic samples contain phenocrysts of clinopyroxene±olivine±plagioclase with clinopyroxene — the dominant phenocryst phase. Subophitic texture, with small plagioclase lathes intergrown with clinopyroxene, was observed in several thin sections of basalts collected from the boreholes DH12 and YM201.

2.1.2 Layered Intrusive Rocks

The Wajilitag layered mafic–ultramafic intrusion, as an important lithological unit of the Tarim LIP, was located in the southeastern part of Bachu County in the western Tarim Basin (Fig. 2.1). It consists mainly of olivine pyroxenite, pyroxenite and gabbro from the lower part to the upper part and shows rhythmic layered structures. Olivine from the mafic–ultramafic rocks has low Fo values [molar $100\times Mg/(Mg+Fe)$] of 67–76 mole%, probably indicating that the Wajilitag magmas have experienced fractional crystallization with an early removal of olivine. Ilmenite exsolution lamellae are present in clinopyroxene. Layered structures and variable mineral compositions of olivine and clinopyroxene of the intrusion suggest that the magmas become more Fe–Ti-rich under extensively fractional crystallization from the evolved magma.

Zhang et al. (2008) divided the complex into four main lithofacies: (1) magnetite–olivine pyroxenite (accounting for ~5% of the outcrop), (2) olivine-bearing pyroxenite (~30%), (3) gabbro (~60%), and (4) syenite and quartz syenite (~5%) after previous petrographic studies (Li et al., 2001), and they described the petrography of those rocks well. Transitional rock types, such as pyroxene-bearing diorite, olivine-bearing gabbro, pyroxene and/or nepheline-bearing syenite, also occurred in this area. All of the rocks are relatively fresh, with only slight metasomatism, as represented by sericite as a replacement for plagioclase. They show a clear medium- to coarse-grained texture and a block structure. The magnetite–olivine pyroxenite lithofacies is made up mainly of olivine (10%–20%), augite (40%–50%), plagioclase (10%–20%), magnetite (10%–30%), and minor amounts of phlogopite (~1%) and apatite (<1%). In augite, the glint texture (where titanium and iron oxide are parallel to the cleavage of augite) is visible. The pyroxenite lithofacies is made up mainly of clinopyroxene (75%–95%), with minor amounts of olivine, plagioclase, and Ti–Fe oxides. Gabbro, the predominant lithofacies of the complex, consists of clinopyroxene (40%–50%), plagioclase (40%–45%), and minor amounts of hornblende, phlogopite, and apatite. Orbicular structures with dimensions of ~50 cm were observed on several outcrops in this lithofacies. The quartz-syenite lithofacies is exposed in a topographic high near the center of the complex. The main minerals are alkaline feldspar (orthoclase, 40%–70%), quartz (10%–30%), hornblende (5%–10%), and brown biotite (1%–2%). The accessory minerals include zircon, apatite, rutile, allanite, and monazite.

On several outcrops, minor nepheline-bearing syenite occurred. The nepheline-bearing syenite is made up of alkaline feldspar (40%–60%), pyroxene (40%–50%), nepheline (1%–5%), and phlogopite (1%–5%) as well as accessory minerals, such as apatite, allanite, xenotime, and baddeleyite.

2.1.3 Breccias

The Wajilitag ultramafic cryptoexplosive breccia (Fig. 2.2D), as one of the most important rock types of the Permian Tarim LIP, is made up of lapillus and groundmass. The size of lapillus ranges from several millimeters to tens of centimeters, and the shape changes from angular, sub-angular to sub-rounded. The components of lapillus are mainly dunite, clinopyroxene–peridotite, olivine pyroxenite, and clinopyroxenite, which have a holocrystalline medium- to coarse-grained texture and a massive structure with no plastic deformation. The groundmass is volcanic lava, which has a porphyritic texture. Most of the phenocrysts are olivine and clinopyroxene, and some are phlogopite and hornblende. The groundmass has a microcrystalline texture, made up of olivine, clinopyroxene, phlogopite and so on. Auxiliary minerals such as apatite, magnetite, ilmenite, chromite, perovskite, zircon and spinel can be seen in the breccias and groundmass. Besides, corundum, rutile, pyrope, anatase and moissanite can also be found in the cryptoexplosive breccia (Wang and Su, 1990; Li et al., 2010).

2.1.4 Picrite and Olivine Pyroxenite

•Picrite

All of the picrite samples recovered from the boreholes in NTU are altered; however, typical petrographical features, such as the porphyritic texture and pseudomorphs after olivine phenocrysts, are well preserved. Petrographically, recognized groups are picrite, basalt (including dolerite), and two groups of rhyolite defined by their compositions. Abundant olivine phenocrysts in the picrites provide evidence of crystal accumulation. Alteration of the rocks is manifested by partial to complete replacement of olivine by iddingsite, plagioclase by sericite, glass by chlorite, infilling of vesicles with calcite and ankerite, and by the presence of veins and patches of chlorite and carbonates. The extent of the alteration of picrites varies from slight to severe for the picritic samples. A detailed petrographical description of all the samples is available in Tian et al. (2010).

• **Olivine pyroxenite**

The Xiaohaizi ultrabasic diked rock intruded into the Silurian to Carboniferous strata (Fig. 2.2B) has a granular texture and a massive structure. Olivine and clinopyroxene are coarse-grained, and curved cracks and marginal corrosion, as well as black ferrous minerals along the cracks, and crystal margin can be seen in the olivine. The gaps are filled with plagioclase and mafic accessory minerals, mainly of magnetite and apatite, are interstitial between olivine and pyroxene crystals. The calculation of CIPW standard minerals shows that there are 4%–15% olivine, 46%–59% augite, 16%–19% plagioclase and 17%–20% accessory minerals in the rock. The appearance of plagioclase may be caused by late-stage cooling crystallization rather than residue of fractional crystallization. In the calculation of the CIPW, the content of FeO is not used, so apart from plagioclase and apatite, the contents of other minerals should be corrected. After the correction, the average modal content of olivine is about 15%, pyroxene 60%, and the accessory minerals less than 10%. According to the Cpx-Opx-Ol diagram for classifying the ultramafic rocks and ultrabasic rocks, combined with observation under the microscope, the ultrabasic vein rock can be named as olivine–ehrwaldite.

The olivine in the Xiaohaizi olivine–ehrwaldite (xhzn2-1) has a high MgO content of 39.87–42.89 wt.%, $Mg/(Mg+Fe^{2+})$ ratios of 0.78–0.81, and the end member component of $Fo_{78-81}Fa_{22-19}$. On the En-Fs-Wo classification diagram, most of the clinopyroxene in the olivine–ehrwaldite is augite, and some points in the area between augite and diopside. Clinopyroxene owns the end member component of $Wo_{37-45}En_{40-53}Fs_{10-20}$, and high $Mg/(Mg+Fe^{2+})$ ratios of 0.70–0.85. Al_2O_3 in the plagioclase is up to 26.47–30.10 wt.%, On the An-Ab-Or triangle classification diagram, plagioclase has an anorthite content of 45–68, belonging to andesine and labradorite.

2.1.5 Syenite Body

The Permian Xiaohaizi quartz syenite body and dikes were developed well in the Xiaohaizi area of Bachu County in the northwestern Tarim Basin (Yang et al., 2006a, b, c, 2007; Sun et al., 2008; Li et al., 2011; Wei and Xu, 2011). Both the Xiaohaizi quartz syenite body and dikes, consisting of the ultra-basic, basic and intermediate-acidic dikes, intruded into the Silurian, Devonian, Carboniferous and Lower Permian strata, of which the latter did not cut into the quartz syenite body. The Xiaohaizi quartz syenite body occurred in several square kilometers. It has a main body of a hornblende syenite and a hornblende syenite was surrounded by the olivine syenite along the northern, western and southern sides with the widths of ca. 100m respectively, and the former was intruded by the latter (Wei and Xu, 2011). Depending on the field relationship between the syenite body and dikes, it can be determined that the emplacement time of the syenite may be later than that of the dikes.

2.1.6 Syenitic Porphyry, Diabase and Bimodal Dikes

Bimodal dikes may provide the key to understanding the rifting environment of the Tarim Block. Bimodal volcanic suites form in various tectonic settings, such intracontinental rifting (Shao et al., 2001; Li X.-H. et al., 2002; Lindline et al., 2004; Vorontsov et al., 2004; Li W.-X. et al., 2005; Shu et al., 2005), back-arc basins (Hochstaedier et al., 1990; West et al., 2004), island arcs (Geist et al., 1995; Qing and Wang, 1999), active continental margins (Donnelly and Rogers, 1980), and rifting of the passive margin (Oberc-Dziedzic et al., 2005). They have also been genetically correlated with mantle plumes (Li X.-H. et al., 2002; Li Z.-X. et al., 2003; Jordan et al., 2004; Li W.-X. et al., 2005). Therefore, the study of bimodal igneous suites can provide important clues to the geodynamics and crust-mantle interaction (Yang et al., 2007). The bimodal dikes of quartz syenite porphyry and diabase occurred in the Xiaohaizi area of Bachu County and their lithology was described as follows.

The massive quartz syenite porphyry shows mostly coarse phenocrysts of K-feldspar with minor plagioclase, together with microlites of K-feldspar (70%–90%) and minor plagioclase. The groundmass K-feldspar (0.5 mm) in this rock is coarser than that in the dike (0.25 mm) by a width of ca. 1.5 m, reflecting their difference in the order of crystallization. Some fine-grained (up to 0.25 mm) anhedral quartz (<5%) commonly occurs interstitially in microlitic K-feldspar/plagioclase. Microlitic plagioclase in some samples has locally preferred orientation and the rocks show a pilotaxitic structure in the ground mass. Accessory minerals are apatite, zircon, titanite and iron oxides.

Diabases occur in direct contact with the quartz syenite porphyry (Fig. 2.2A) and consist of phenocrysts of plagioclase together with a groundmass assemblage of microlitic plagioclase, clinopyroxene and hornblende. The rock shows mineral assemblage and igneous textures compared with other diabases in the Bachu area.

2.1.7 Felsic Volcanic Rocks

Felsic volcanic rocks in the Tarim LIP have a minor distribution size of 5 km^2

compared with huge continental flood basalts in the Tarim Basin, and occurred in the NTU and Kuche Depression of the North Tarim Basin (see Fig. 1.9). They consist mainly of rhyolite, dacite, and ignimbrite (Chen et al., 1998; Tian et al., 2010; Liu et al., 2014), and occurred as interlayered with basalts commonly in the field sections (Yu et al., 2009; Shangguan et al., 2012), deduced as products of the Tarim LIP. Some thick felsic volcanic rock layers are overlapped in the layered basalts in some boreholes except for those interlayered with basalts in boreholes (Tian et al., 2010; Yu et al., 2011; Yang et al., 2013), indicating at least two-episodes of felsic volcanic activities in the Tarim LIP.

The above arguments are clarified by the chronological results. According to the zircon U–Pb dating from felsic volcanic rocks obtained (Liu et al., 2014), they have formation ages of concentrated two episodes, the first-stage ages of 291–287 Ma having the same ages as the Tarim basalts, and the secondary-stage ages of 283–272 Ma having the same ages as most of the Tarim intrusive rocks (seen in Section 2.3) as the contemporary products.

2.2 Spatial Distribution of the Tarim LIP

Huge volumes of Permian basalts and related igneous rocks (291–274 Ma) are widely distributed in the Tarim Basin. They mainly occur in the west of the NTU, in the Awati Depression, the Keping Rise, and the Bachu Rise in the center-west part of the Central Rise and in parts of Southwest Tarim Basin (Fig. 2.1). The igneous rock consists mainly of basalts, diabases, basaltic andesites, ultramafic rocks and syenites. Field contact relationships between Permian igneous rocks (basalt, basaltic andesite, olivine pyroxenite, syenite, bimodal dikes, diorite and diabase) and sedimentary rocks have allowed an accurate limit of time relationships between rock units, which were previously poorly limited due to the absence of a biostratigraphy. Thus, this can potentially also provide insights into the early rates of some geological processes and secular changes in those rates over the course of the history of the Earth (Blake, 2004).

2.2.1 Basalts

The Permian basalts in the Tarim Basin are widely distributed in the western and central parts of the Basin (Fig. 2.1). Yang et al. (2007a) reported that the thickness of basalt lavas ranges from tens to hundreds of meters, and the spatial distribution area is estimated to be over 2.5×10^5 km^2, the latter of which is comparable to the Emeishan LIP (Chung and Jahn, 1995; Xu et al., 2001; Yu et al., 2009).

The volcano-sedimentary sequence in the Basin can be divided into the Kupukuziman Formation at the bottom and the Kaipaizileike Formation at the top in the Lower Permian stratum. They are mainly exposed in the Yingan, Kaipaizileike and Sishichang areas. Basalts with a columnar joint structure are present in the Kaipaizileike area. In the Keping area (Fig. 2.2G), 2 basaltic lava flows (Group 1b basalts) occur within the Kupukuziman Formation and 6 basaltic lava flows (Group 1a basalts) occur within the Kaipaizileike Formation. They are inter-bedded with Permian fluvial sedimentary strata. The lava flows are composed of amygdale basalt, olivine basalt, basonitoid and dolerite. Interlayered sediments containing tuffs of grey to green colors are inter-layered with some of the basaltic lavas.

The 2 basalt sequences are prominent landscape markers and each is laterally traceable on a regional scale in the field and in aerial images (see Fig. 1.7). The lower basalt sequence encompasses 2 basalt units in the Yingan, Kaipaizileike and Sishichang sections, while the upper basalt sequence may include 6 basalt units, although all 6 are only exposed in the Yingan area, with the Kaipaizileike and Sishichang sections truncated from the action of erosion (Yu et al., 2011).

Evidence of this link regarding the same ages, between 290–285 Ma, is also supported by a biostratigraphical correlation between the basalt-bearing Qipan and Kupukuziman formations on the basis of the common occurrence of the microfloral *Potonieisporites–Vestigisporites* assemblage (PVA) (Chen and Shi, 2003). Comparable basalts found in the boreholes in the central Tarim region are usually assigned to the volcanic formation (Chen and Shi, 2003). No radiometric ages are available from these basalts to date, nor has this formation yielded any fossils, leaving a direct stratigraphic correlation of the subsurface basalts uncertain. The volcanic formation is 290–340 m thick, conformably overlies the lower clastic formation, and is in turn overlain by the upper clastic formation (Chen and Shi, 2003). The lower clastic formation ranges from 1 to 225 m in thickness, and is characterized by a microfloral PPA (*Potonieisporites–Pityosporites* assemblage), which is correlative to the PVA recognized from the basalt-bearing strata of both the Keping and the southwestern Tarim areas (Chen and Shi, 2003). The upper clastic formation has been truncated by the action of erosion and unconformably overlain by the Neogene sediments. Its depositional thickness is therefore unlimited, but the preserved section in various boreholes is up to 800 m thick. This formation contains the ostracod *Darwinula–Panxiania* assemblage (DPA), which also occurs in the Shajingzi Formation capping the Kaipaizileike basalts in the Keping area (Chen and Shi, 2003).The Lower Clastic Formation is thus equivalent to the pre-basaltic lower Kupukuziman Formation, while the Upper Clastic Formation correlates to the post-basaltic Shajingzi Formation. It follows that, although supporting a general correlation between the Volcanic Formation and the Keping basalts, in the absence of a direct chronometric constraint, a specific correlation (and accompanying limit on its areal extent) cannot be drawn to any of the specific flows from the Keping region, or even to either or both of the two eruptive stages. Direct dating of the Volcanic Formation would thus represent an important additional control on the geographical distribution and progressive evolution of the Permian Tarim basalts.

Detailed field work has led to recognition of 8 basalt units (BUs) in the Permian sequences exposed in the Keping area of the northern Tarim Basin. These basalt units are interbedded with terrestrial clastics, and thus represent eight discrete eruptive episodes. The basalt sequence is divided into two major eruption cycles, with an earlier cycle comprising 2 BUs within the uppermost Kupukuziman Formation, separated by ~1000 m of clastic sediments from an upper eruptive cycle represented by a further 6 BUs in the upper part of the Kaipaizileike Formation (Yu et al., 2011).

The Permian basalts crop out largely in the western part of the basin, with significant exposures found in the Xiahenan and Keping areas (Fig. 2.1). In the latter, the complete Permian basalt succession is exposed in the continuous Yingan, Kaipaizileike and Sishichang sections (Yu et al., 2011).

Although less well exposed than their counterparts around Keping, Permian basalts in the Xiahenan area form spectacular geomorphic features parallel to stratigraphic beds. Further minor exposures of Permian basalts are also recognized in southwestern Tarim, although these basalts are restricted to the Qipan and Damusi sections, where they form beds that are 2–3 m thick in the upper part of the Qipan Formation (Chen and Shi, 2003; Li et al., 2008). In the central Tarim desert area, although not occurring in outcrop, further occurrences of Permian basalts are widely noted in boreholes (Fig. 2.3), and the overall distribution of these distinctive basalts in the Tarim Basin may extend across to 2.5×10^5 km^2 (Chen et al., 1997; Yang et al., 1996, 2005, 2007; Chen et al., 2006). The Permian basalt successions are up to 500 m thick in the Yingan section of the Keping area. Comparable thicknesses of 549 m are recognized in the subsurface record from borehole TZ22 on the Tazhong Uplift, with 478 m and 443 m in boreholes He4 and Shan-1 on the Bachu Uplift, and 372 m, 442 m and 207 m found in borehole YM5 on the Tabei Uplift and boreholes MX2 and HD5 in the Manjaer Depression, respectively (Chen et al., 2006; Yu et al., 2011).

●Yingan section

The Yingan section is located in the northwestern part of the Tarim Basin and a field occurrence of 180°∠44° in the surrounding sedimental strata. The basalt layers have a total thickness of ca. 530 m in this section except for small amounts of intercalated Permian sedimentary layers composed of mudstones, siltstones, and mud-bearing limestones.

Spectacular Permian successions are exposed in the Yingan section, which is located near Yingan Village, about 45 km northeast of Keping, in southern Xinjiang. Two pronounced basalt sequences stand out dramatically in a section dominated by pale silici clastic sediments assigned to the Kupukuziman (P_{2kk}) and the Kaipaizileike (P_{2kp}) formations respectively (Chen and Shi, 2003). The Kupukuziman Formation consists of alternating reddish mudstone and siltstone, interbedded with numerous conglomerate layers in its lower and middle parts. The lower basalt sequence occurs in the upper part of the Kupukuziman Formation and includes 2 discrete basalt units (BUs). The oldest, BU1 is ca. 10 m thick, with BU2 approximately double this thickness. BU2 is separated from BU1 by several meters thick of intervening reddish siltstone, and is itself overlain by a 10 m thick greyish-green tuff bed, which is succeeded by a non-marine limestone bed containing freshwater fossils, which marks the top of the Kupukuziman Formation. The overlying Kaipaizileike Formation is made up of alternating siltstone and mudstone of terrestrial facies. The lower ca. 800 m of this formation lack igneous material, but 6 prominent basaltic lavas are interbedded with the upper layers of the sequence. BUs 3–5 are each ca. 20–40 m thick, with BU6 and BU8 on the order of doubling this, at approximately 72 m and 60 m, respectively. BU7, also up to 70 m thick, stands out from the other basaltic horizons as it is mainly made up of andesitic basalt.

•Kaipaizileike section

The Kaipaizileike section is situated approximately 20 km northeast of the Yingan section (Fig. 2.2E). As at the Yingan section, the Permian sequences exposed here encompass the Kupukuziman and Kaipaizileike formations, each bearing a set of basalt units. The lower basalt sequence preserved here in the Kupukuziman Formation again consists of 2 distinct basalt units (BU1 and BU2), separated by ca. 20 m of terrestrial sediments. Only 5 basaltic lavas are recognized within the overlying Kaipaizileike Formation in this section, however, with BU8 not cropping out, probably due to erosional truncation of the sequence, with a profound unconformity present locally between the Permian beds and the overlying Neogene sediments (Yu et al., 2011). Both BUs 3 and 4 are about 70 m thick in the Kaipaizileike section, and are separated from the Kupukuziman basalt sequence by approximately 1000 m of terrestrial sediments. BUs 5 and 6 are 50 m and 70 m thick, respectively, and BU7 comprises up to 75 m of andesitic basalt, lithologically correlative with the equivalent basalt unit (BU7) identified in the Yingan section. This basalt unit is followed by a 100-m-thick succession of reddish-purple terrestrial silici clastics, which are themselves overlain unconformably by upper Neogene yellow-brown siltstones interbedded with conglomerates and sandstones.

•Sishichang section

The Sishichang section is approximately 12 km northeast of the Kaipaizileike section. Here, the exposed Permian sequence includes only the Kupukuziman Formation. This is unconformably overlain by the upper Neogene, with the Kaipaizileike Formation and the upper basalt sequence absent. Like its counterpart exposed at both the Yingan and Kaipaizileike sections, the Kupukuziman Formation is here characterized by reddish terrestrial mudstone, siltstone and conglomerates capped by 2 basalt layers, 10 m and 23 m thick, and separated by ca. 30 m of sediments. These basalt layers are correlated to the lower basalt sequences BUs 1 and 2 seen in the Yingan and Kaipaizileike sections.

The samples analyzed were collected from 5 field sections of the Yingan, Sishichang, Xiahenan, Wajilitag, and Xiaohaizi areas, and the boreholes of He4, H1, YT6, SL1, YM5 and YM8. In addition to the surface exposures of the flood-volcanics, several boreholes drilled for oil and gas exploration have sampled basaltic rocks at the depths of 810 m to 5900 m in the central and northern parts of the Tarim Basin.

The various units of basaltic rocks identified are coeval and chemically similar (Li Z.-L. et al., 2008; 2012; Tian et al., 2010). Their geological features are given in Zhang et al. (2003), Jia (1997), Yang et al. (2005) and Li et al. (2008). In general, the basalts have been classified into various compositional groups. The Group 1a and Group 1b basalts are aphyric and locally hyaloclastitic, occasionally vesicular/amygdaloidal, and sometimes with rare phenocrysts of plagioclase (<3%). They show interlayering of porphyritic and aphyric varieties. The groundmass is made up of plagioclase (30%–50%), basaltic glass (30%–45%) and Fe–Ti oxides (5%–10%). The basalts belonging to compositional Group 2 basalts are augite-phyric (1%–2%) and plagioclase-phyric (<2%), with minor olivine, and occasional lavas are aphyric with rounded vesicles. Some lavas are more strongly porphyritic, with 2%–10% plagioclase phenocrysts and sparse phenocrysts of olivine (1%–2%) and augite (1%–3%). Groundmass textures vary from interstitial, through microlitic, to devitrified glassy, and the crystalline groundmasses are typically made up of plagioclase (30%–45%), clinopyroxene (30%–45%), olivine (<45%) and Fe–Ti oxides (5%–10%). The magmatic stratigraphy of the Yingan section of Keping area was described in detail in Yu et al. (2011) and Li Y.-Q. et al. (2012). The Yingan section in Keping County is one of the best Permian basalt outcrops in the study area and was hence selected for Hf-isotopic investigation in this study. The stratigraphy and geochemistry, including PGE were described by Yu et al. (2011) and Li Y.-Q. et al. (2012), where the 8 basaltic lava flow units of a total thickness of over 500 m were recorded. Among these, 2 units are exposed in the Kupukuziman Formation and 6 units in the Kaipaizileike Formation. Individual basaltic units range from 10 to 70 m in thickness, with intercalations of terrestrial sediments such as alternating reddish mudstone and siltstone, volcanic breccia, tuff and non-marine limestone (Yu et al., 2011). The basalts from the middle and upper parts of the 5th unit are exceptional in that they carry abundant plagioclase phenocrysts (ca. 40 modal%).

●Damusi section

A suite of basalt layers, found in the Damusi section in Zepu County of the southwestern part of the Tarim basin (Fig. 2.1 and Fig. 2.2F), are interlayered in the middle part of the Early Permian Qipan Formation with an orientation (strike/dip) of 220°/50° and a thickness of approximately 42 m. A more detailed description can be seen in Zhang et al. (2003).

•Summary

The spatial distribution of the Tarim LIP from the spatial section lines shows the stratigraphic correlation among basaltic lava, tuff, interlayered mudstone, siltstone and sandstone and the thicknesses of the basaltic lavas from different field sections and boreholes, and the basalts from the Kupukuziman and Kaipaizileike formations were subdivided in the borehole sections. This indicates that the basaltic lavas were widely distributed in the Tarim Basin.

The Kupukuziman basalts are inter-layered with tuffs and can be subdivided into 2 separate lavas (Fig. 2.2G). Both lavas are composed of amygdale basalt and olivine basalt intercalated with tuff layers with a total thickness ranging from 46.4 m to 86.5 m in the Yingan and Sishichang sections, respectively. The basalts in the Kaipaizileike section are inter-bedded with tuff and fine-grained clasts; they collectively have a total thickness of 300 m. The associated sedimentary strata in the Yingan area have a strike/dip of 180°/44°. The total thickness of basaltic lavas excluding inter-layered mudstones, siltstones and limestones in the Yingan section is 370 m (Yu et al., 2011).

In the Damusi section located in Yecheng County in the southwestern Tarim Basin (Fig. 2.1), massive basaltic lavas of ~42 m in total thickness are present within the middle part of the Qipan Formation in the Early Permian period (Fig. 2.2F). The sedimentary strata of the Qipan Formation have a strike/dip of 220°/50° (Li et al., 2008).

Permian basalts in the Tarim Basin occur in the outcrops around the basin (the Keping and Taxinan areas) and have also been identified from drilled cores within the basin (Jiang et al., 2004b, 2006; Yang et al., 2005, 2006a; Li et al., 2008; Yu et al., 2009, 2011). The basalts have been recognized in the Kupukuziman Formation in the lower part and the Kaipaizileike Formation in the upper part of the Early Permian (Jiang et al. 2006; Yu et al., 2011). Li et al. (2008) argued that the basalts exposed in the lower Permian Qipan Formation of the southwestern Tarim Basin have geochemical affinity with those from the Keping area, and correspond to those from the Kupukuziman Formation in the Keping area. The data from the boreholes and 3D seismic cross-sections show a thickness of ca. 2.5 km for the basaltic sequence and a total volume of 7.5×10^4 km^3 (Zhou et al., 2009; Tian et al., 2010; Yu et al., 2011). The Permian magmatic event is also manifested in eastern Kazakhstan, Salair, western Mongolia and large regions to the north between the Siberian Craton and Altay orogenic belt (Pirajno et al., 2008; Ao et al., 2010).

Two thick basaltic lava sequences are present in the Kupukuziman Formation of the lower Permian area in the Sishichang section. The first basaltic sequence occurs above the Upper Carboniferous to lower Permian sedimentary strata. Tuff and limestone with fossils were in-bedded within the basaltic sequence in the Yingan section (Fig. 2.2G). The Qipan Formation corresponds to the lower part of the Kupukuziman Formation in the Keping area. Diabases in the region were emplaced after a basaltic eruption because they cut across the entire Kupukuziman Formation in the Yingan section.

The large volume of eruptions from the Keping area in the southwestern part and in the central part of the Tarim Basin, combined with the enriched geochemical signature, support the hypothesis that Tarim basalts were derived from a plume volcanism, either from the plume itself or from the melting of the enriched lithospheric mantle that was heated and uplifted by an ascending mantle plume (Li et al., 2008).

The field outcrop and borehole observations and descriptions display large-scale Permian igneous events in Tarim and its surrounding regions, particularly, the large volume of the Permian mafic igneous rock occurred. Recent geochronological and stratigraphic (especially detailed paleontology) studies on the Permian basalts in Tarim and its marginal areas show that a large volume of basalts erupted between 291 and 284 Ma (Chen et al. 1997, 2006; Jia et al., 2004; Li et al., 2011). According to a geophysical exploration and borehole data (Jia, 1997; Chen et al., 2006), the coverage area of the Permian basalts (including related tuff and tuff-bearing rocks) in Tarim is ca. 2.5×10^5 km^2. Borehole and section data (Jia et al., 2004; Chen et al., 2006) indicated that the thickness of the basalts varies between ~100 and ~800 m, with an estimated average thickness of ca. 300 m based on the published borehole and section data (Jia et al., 2004; Jiang et al., 2004b). Thus, the volume of the Permian basalts in Tarim is estimated at ca. 7.5×10^4 km^3. If the coeval basalts in the Tuha and Sangtanghu basins, north of Tianshan (Zhou et al., 2006) and the widely distributed Permian mafic dikes in Tarim (Jiang et al., 2004b) are included, the total volume of the basaltic rocks could reach up to 1×10^5 km^3 (Zhang et al., 2008).

Li et al. (2014) studied the stratigraphy and sedimentology of 85 boreholes/outcrops and 2 seismic profiles to better understand the crustal uplift/erosion of the Tarim LIP related to a mantle plume. The results indicate a regional unconformity and onlap contact between the earliest Permian and Carboniferous strata, with denudation of the Carboniferous stratum and the deposition of the earliest Permian strata. The denudation of the Carboniferous strata has been divided into 3 types of erosion: deep, partial and weak, with a deep erosion zone located in the northern Tarim region (east of Keping and west of northern Tarim) which gradually weakens towards the south. Above the unconformity, the Early Permian pre-eruption deposit is thinnest (<100 m) in the northern Tarim region, and thickens gradually to the south (>300 m), which is consistent with the spatial distribution of the Late Carboniferous erosion. This unconformity represents a spatial distribution of denudation forced by doming uplift/erosion at the end of the Carboniferous period, with the minimum estimation of its vertical and lateral extent of 887 m and ~300 km, respectively, which are comparable to those of other mantle-plume-generated LIPs. Both the Late Carboniferous erosion and Early Permian sedimentary records indicate a crustal uplift event at ~300 Ma. The uplift was short-term and approximately dome-shaped, occurring immediately before the Tarim LIP (~290 Ma). The spatial distribution of denudation suggests that the regions of Northern Tarim, or further north, represent the center of the potential plume head.

2.2.2 Layered Intrusive Rocks

●**General intrusive rocks**

The intrusive rocks from the Bachu area (including Xiaohaizi and Wajilitag) include diabases, layered mafic–ultramafic intrusions, mica-olivine pyroxenite breccia pipes and ultramafic dikes, quartz syenites, and quartz syenite porphyry. In the Bachu area, where the dike association is well developed, it is seen that syenite batholith, ultramafic rocks, and quartz syenitic dike intrude into the Silurian, Devonian, and Carboniferous–Lower Permian strata (Fig. 2.2A, B; Yang et al., 2007).

●**Intrusive rocks**

The Wajilitag mafic–ultramafic complex, located in the Bachu area of the Tarim

Basin, mainly includes olivine pyroxenite, pyroxenite and gabbro and diabase, and hosts Fe–Ti minerals. They intruded into the Devonian reddish sandstones. Irregular intrusive rocks contact with the Devonian sandstones and siltstones commonly. The intrusions can be further divided into layered mafic–ultramafic intrusion (olivine pyroxenite), breccia pipe ultramafic rock and diabase. The layering structures are sub-horizontal with a gentle dipping towards the southeast. The most common intrusive rock types include pyroxene peridotite, (olivine) pyroxenite, gabbro, diorite and syenite (Li et al., 2001). Six explosive mica-olivine pyroxenite pipes and 32 mafic dikes have been found in the region (Wang and Su, 1987). The breccia pipes and mafic dikes intruded the Upper Devonian continental clasts (Li et al., 2001). The dikes are mainly diabases and lamprophyric dikes containing Mg–Na amphibole (Li et al., 2001).

2.2.3 Breccias

The Wajilitag area is located in the southeastern part from Bachu County at a distance of 40 km. A basic–ultrabasic complex pluton covers an area of about 10 km^2 and is made up of a layered rock body, cryptoexplosive breccia pipes and a late-stage diabase dike. The layered rock body is made up mainly of pyroxenolite and gabbro assemblages, and some transitional rock types. The cryptoexplosive breccia pipe is made up mainly of ultramafic cryptoexplosive breccia, and 6 cryptoexplosive breccia pipes (Wang and Su, 1987), with a negative form of landform. Wang and Su (1987) regarded them as volcanic vent facies and crater facies from a field observation. Li et al. (2001) considered them as breccia hornblende olivine pyroxenolite for a large number of pyroxene breccias and clinopyroxene xenolith, and they regarded the breccia hornblende olivine pyroxenolite as well as the late-stage vein as products of the same period of magmatic activity that occurred after the solidification of the layered intrusive rocks. The late-stage diabase dikes are distributed widely and occur as a radial pattern within and around the layered intrusive complex in the area.

The kimberlitic cluster trends NWW–SEE within an area of 5 km^2 and is deeply weathered at the surface to a greenish clay. These kimberlitic pipes and dikes intruded the flat-lying and metamorphosed continental clastic sequences of the upper Devonian Keziletag and Yimugangtawu formations, and were in turn cut by late dolerite dikes. Among these pipes with several micro diamonds separated so far, it is the most important diamondiferous pipe in the area and is an oval-shaped body with dimensions of ca. 180 m×100 m (Su, 1991; Bao et al., 2009). The Wajilitag kimberlitic intrusions are spatially associated with the Wajilitag Early Permian Fe–Ti oxide ore-bearing ultramafic–mafic-syenitic intrusion that is made up of clinopyroxenite, gabbro and syenite and was emplaced at ca. 274 Ma (Zhang et al., 2008), but no direct intrusive relationship among these rock units has been found in the region (Li et al., 2011). Moreover, numerous felsic and mafic–ultramafic dikes (including some carbonatites) with variable strikes intruded either the upper Devonian strata or the various intrusions in the area (Zhou et al., 2009; Zhang C.-L. et al., 2010; Zhang D.-Y. et al., 2013).

2.2.4 Picrite and Olivine Pyroxenite

●Picrite

The Permian igneous rocks, recovered from industrial (oil) drilling from the NTU, confirmed a picrite–basalt–rhyolite suite in the NTU and were recently studied from 13 boreholes at depths between 5166 and 6333 m (Tian et al., 2010). They underlie a Mesozoic sedimentary succession and consist mainly of basalts, dolerites and rhyolites, with minor picrites. An amygdaloidal structure is common in basaltic samples, implying a subaerial eruption. Picrites are found in the YT1 and YH5 boreholes. Rhyolites, in variable amounts, were mainly recovered in the western part of the NTU. The rough proportions of felsic rock and basalt intersected in each of the boreholes are shown by pie charts at the borehole locations. Two deep boreholes, YM16 and YM30, both recovered more than 1000 m of igneous rocks. Rhyolite is dominant in YM16 with minor ignimbrite, whereas basalt, interbedded with minor rhyolite layers, is the dominant rock type in YM30. However, the bottoms of the volcanic sequences were not reached in either borehole and our estimate of the actual thickness of the volcanic sequences is based on high-quality three-dimensional (3D) seismic data. A 3D seismic cross-section, through the prophyry-dominated sequence intersected in YM16, shows a dome-like structure that extends deep into the crust (as illustrated in Fig. 2 of Tian et al., 2010). A 3D seismic cross-section passing YM30 and YM8, about 10 km in horizontal length, shows that the basaltic sequence is well layered. The seismological thickness of the basalt layer exceeds 1200 m below YM30 and exceeds ~2500 m below YM8. The upper part of the basaltic layer around borehole YM30 is eroded and angular-unconformably covered by later sediments. So the original basalt layer was thicker than that intersected by drilling. Another

3D seismic cross-section passing YM8 and YM5, about 30 km long, shows that the thickness of the basalt layer is well in excess of 2500 m. The distribution of the picrite layers in YT1 and YH5 is not well limited, because there is no local 3D seismic data (Tian et al., 2010).

- **Olivine pyroxenite and diabase dike**

The olivine pyroxenite near the southern part of the Xiaohaizi reservoir in the Bachu area occurs as dikes intruding into the Carboniferous sedimentary rocks. In the Xiaohaizi area, an olivine pyroxenite dike occurs sub-vertically (with a dipping angle of ~85°) and has a striking of 83°. It is ~2 m in thickness and is sub-parallel to a fine-grained diabase. The distance between them is only 0.5 m. A thin baked (contact metamorphism) zone is observed along the contacts of the diabase with country rocks, such as siltstone.

2.2.5 Syenitic Porphyry, Diabase and Bimodal Dikes

The Permian Xiaohaizi syenite body and dikes are well-developed in the Xiaohaizi and Wajilitag areas in Bachu County in the northwestern Tarim Basin. Near the Xiaohaizi water reservoir in the Bachu area, a syenite body with surface exposure of several square kilometers is present. The syenitic rocks of late-stage magmatism of the Tarim LIP are sporadically distributed and co-exist with mafic rocks in the Tarim LIP, with the features of bimodal volcanism. It consists of a purple to pale-grey unit in the center and dark-green unit on the margin. The syenite batholith intruded into the Silurian, Devonian, and Carboniferous–Lower Permian strata. Mafic dikes within the syenite intrusion are rare and have different orientations compared to abundant mafic dikes (referred to as a dike swarm here) in the sedimentary country rocks of the syenite intrusion (Yang et al., 2006b, 2007). There are also some nepheline-bearing syenites outcropped on the top of the intrusion, and widespread mafic dikes cut through the intrusion and surrounding sedimentary strata in the Wajilitag area (Zou et al., 2015).

The mafic dike swarm has an overall striking of N70° and sub-vertical dipping. A dike swarm consists of olivine pyroxenite dikes, diabases and quartz syenitic porphyry dikes with an overall trending of NW–NNW and a high dipping angle of >60° is present in the basin, especially in the Bachu area. Individual dikes vary from several tens of centimeters to several meters in width. They cut the Permian and/or Carboniferous sedimentary strata in the basin.

The diabase dikes intruded into the Devonian to Silurian sedimentary sequences in the Xiaohaizi and Wajilitag areas in the Bachu area. In the Bachu area where the dikes are most concentrated, the interval between 2 dikes varies from 1 to 2 m. The diabases in this area can be divided into 3 subgroups based on different orientations: (1) 350°–360° striking dikes with sub-vertical dipping, (2) 310° striking dikes with a sub-vertical dip, and (3) sub-horizontal sills sub-parallel to the sedimentary strata. In each sub-group an individual dike of different compositions occurs sub-parallel. The petrographic features of diabases can be seen in Yang et al. (2007), Zhang et al. (2008) and Li et al. (2011).

The restoration of the geometry of the mafic dikes indicates that the Permian mafic dike swarms in Keping may have extended as far as 61–69 km along a primary orientation of about N320°W, and the flood basalts of the Tarim LIP may have extended to Keping so that the areal extent of the Tarim LIP could be enlarged by about 1.2×10^4 km^2 compared with that originally reported (about 2.5×10^5 km^2; Chen et al., 2014). The geometric features of the mafic dike swarms in Keping and its adjacent areas are different from those of the giant radiating dike swarms due to radial fractures associated with domal uplift. The mafic dike swarms and sills in the Tarim LIP make up the plumbing system of the mantle plume. Accompanied by the upwelling of the mantle plume, many eruptive centers and regional dikes are generated in Beishan, East Tianshan and in the Tarim Basin. The flood basalts in Keping are fissure-type, which may have been fed by magma conduits from the plume center; whereas in Bachu, the central volcanic eruption is dominant.

Field occurrence of dikes in the Bachu area of NW Tarim Basin shows that the dike association is well developed with diabases, ultramafic rocks, and quartz syenitic dikes. The majority of the dikes show a NW–NNW trend with vertical dips and formed during the Early Permian period. In one of the outcrops, a purple quartz syenitic porphyry dike with a width of approximately 1–2 m occurs in direct contact with a dark-bluish diabase dike in the Shuigongtuan area. Both of the dikes show a steep dip of 55°/78°. The dike rocks are surrounded by the Devonian strata (Fig. 2.2A). The diabase dike swarms in the western Tarim Basin intruded into the sedimentary strata during the Early Permian period; however, they did not intrude into the Kupukuziman Formation with basaltic layers, indicating that the diabase formed in the Early Permian period and were earlier than those basalts of the Kupukuziman Formation.

The occurrence of a large number of diabase and basalts, combined with the host Late Paleozoic sedimentary sequences in the Tarim Basin, are indicative of widespread tectono–thermal event associated with crustal extension during the Early Permian period (291–274 Ma).

2.2.6 Summary

In the Wajilitag area, a mafic dike cut the breccia mica-olivine pyroxenite pipes, indicating that the pyroxenite pipes formed before the mafic dike. In the Xiaohaizi area, both quartz syenite porphyry and diabase cut the Silurian and Devonian strata, suggesting that both types of intrusive bodies formed after the eruption of the basalt lavas that are in-bedded in the Kupukuziman and Kaipaizileike formations. The occurrences of mafic dikes and ultramafic intrusive bodies that cut the Silurian–Devonian sedimentary strata in the Xiaohaizi area, combined with the spatial distribution of diabase dikes in the Keping area, suggest that the mafic and ultramafic intrusive rocks formed after the basaltic eruption in the region. The stratigraphic positions of the Permian basaltic lavas in the Tarim Basin are illustrated in Fig. 2.3A and Fig. 2.3B. The two spatial section lines control the whole area of the Tarim LIP distribution, including the field sections such of the Yingan section and the YN1, YM4, AM1, MX2, H4, H3, M5 and QP boreholes, and the P1, H4, BD2, TZ22 and TZ4 boreholes along the NNE–SSW and SEE-NWW directions, respectively. The basalts were distributed in different depths of the boreholes, even arrived at the depth of 4930 m from the surface. The contact relationship between basalts, tuff and sedimentary rocks in the Tarim Basin was also shown. The lines displayed the stratigraphic units and thicknesses of the Permian basalts. The Permian strata were underlay by the Carboniferous strata and covered by the Triassic or much younger strata. There is basalt and tuff in the YN1 borehole, basalt and interlayered mudstone in the YM4 borehole, basalt, tuff and interlayered mudstone in the AM1 borehole, thick basalt and tuff in the MX2 borehole, thick basalt with tuff and mudstone in the H4 and H3 boreholes, basalt with tuff and mudstone in the M5 borehole and thin basalt in QP in the A-A' line. There is a thick basalt with interlayered sandstone in the Yingan section, a thin basalt in the P1 and TZ4 boreholes, a thick basalt with tuff and mudstone in the H4 and TZ22 boreholes, and a basalt, tuff and mudstone in the BD2 borehole in the B-B' line. Some columnar sections from the H4 and TZ22 boreholes of the section lines of A-A' and B-B' selected have the thicknesses of 460 m and 530 m, respectively.

Furthermore, according to the volcano-sedimentary sequences (including the tuffs), the basalt lavas of the Kupukuziman Formation in the lower part and the Kaipaizileike Formation in the upper part was subdivided by the dashed curve lines in Fig. 3, indicating a possible spatial distribution relationship from the field to the boreholes. These data support large-volume spatial distribution and multiple sub-stage eruptions for the Tarim LIP during the Early Permian.

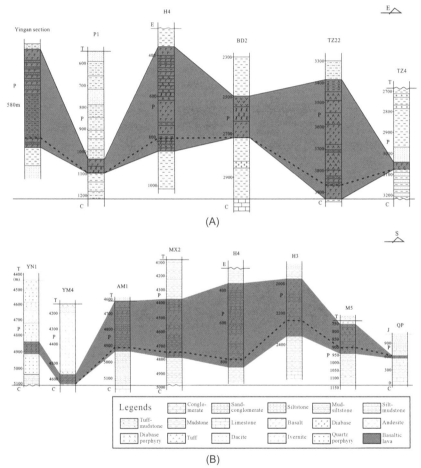

Fig. 2.3 Spatial section lines connected representative field sections and boreholes and displayed the stratigraphic positions and thicknesses of the Permian basalts and contact relationship between basalts, tuff and sedimentary rocks in the Tarim Basin. Note: Dashed curve lines to divide the basalt lava of the Kupukuziman Formation in the lower part and the Kaipaizileike Formation in the upper part according to the relationships between sedimentary strata (including the tuff) and basalt lavas. P: Permian; T: Triassic; C: Carboniferous; J: Jurassic; E: Eocene; and 4400 m: depth of the borehole (Li et al., 2011)

An intrusive relationship between diabases and the Kupukuziman and Kaipaizileike formations basalts in the Keping area suggests that the diabases formed after the eruption of the basalts (Fig. 2 of Li et al., 2011). No direct intrusive relationship is found between the breccia-mica-olivine pyroxenite pipe and a layered intrusion that is made up of pyroxene peridotite, pyroxenite, gabbro and syenite in the Wajilitag area, but rock fragments from the intrusion are found in the breccia pipe. This suggests that the breccia pipe formed later than the layered intrusion. The breccia pipe was cut by a diabase dike. The field relationship described above suggests that the layered intrusion formed before the breccia pipe and the diabase dike formed after the breccia pipe. In the Xiaohaizi area, syenitic body intruded into the sedimentary strata inter-bedded with basaltic lavas, indicating that the basalts erupted before the emplacement of the quartz syenite body. In summary, the temporal evolution of the Tarim LIP from older to younger is from basalts to diabase and finally to syenite (Fig. 2.4).

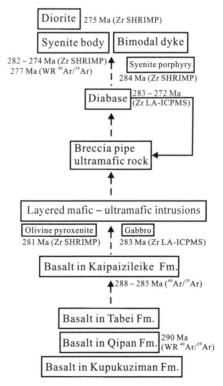

Fig. 2.4 Temporal evolution of the different rock units in the Tarim LIP (Modified from Li et al., 2011). Dashed arrow: temporal order; Solid arrow: intruded contact relationship

Coeval basalts and mafic–ultramafic rocks dispersed in the surrounding area of the Tarim Basin (e.g., Tianshan Mountain, Junggar and Tuha basins) are also a part of the Tarim LIP as argued recently by researchers (e.g., Zhang C.-L. et al., 2008, 2010a; Zhou et al., 2009; Zhang Y.-T. et al., 2010; Qin et al., 2011). Furthermore, Pirajno et al. (2009) suggested that this event might be manifested in eastern Kazakhstan, Salair, western Mongolia and large regions to the north between the Siberian Craton and the Altay as well (Pirajno et al., 2009).

2.3 Time Sequence of the Tarim LIP

The Tarim LIP of the Permian age in northwestern China was well surveyed and studied by geologists in the last few decades (Yang et al., 1996, 2005, 2006a, b, 2007a, b; Chen et al., 1997, 1998, 2006; Jiang et al., 2004a, b, c, 2006; Li et al., 2008, 2011; Zhang CL et al., 2008, 2010; Zhang X.-T. et al., 2010; Pirajno et al., 2009; Zhou et al., 2009; Tian et al., 2010; Yu et al., 2011). The Tarim LIP consists of voluminous continental flood basalts formed during the Early Permian age and a number of coeval mafic–ultramafic intrusions and associated Fe–Ti–V oxide deposits (Jiang et al., 2004b, 2006; Li et al., 2008; Zhang C.-L. et al., 2008, 2010; Pirajno et al., 2009; Zhou et al., 2009; Tian et al., 2010; Yu et al., 2011), similar to the Late Permian Emeishan LIP in southwestern China (Zhong et al., 2002; Zhou et al., 2005, 2008; Wang et al., 2008; Pang et al., 2008a, b, 2009).

Recently, highly-precise geochronological dating (such as U–Pb zircon SIMS, SHRIMP and Laser Ablation Inductively Coupled Plasma-Mass Spectrometers (LA-ICP-MS) and $^{40}Ar/^{39}Ar$ isotopic dating) revolutionized the study of mafic and felsic igneous rocks within continent, and continental margin and extensive setting in the Precambrian and Phanerozoic periods, and provided a precise time-scale limit to some important geological magmatic events throughout the history of the Earth (e.g., Wilde et al., 2003; Blake, 2004; He et al., 2007; Fekiacova et al., 2007; Halls et al., 2008; Kryza and Zalasiewicz, 2008; Xu et al., 2008; McNutt and Nishimura, 2008). The recently precised zircon U–Pb and whole-rock $^{40}Ar/^{39}Ar$ dating from Permian basalts and diabases as well as syenites in the northwestern and southwestern Tarim Basin can provide a good age limit to the tectonothermal event (Table 2.1).

Table 2.1 Geochronological data of the Permian Tarim igneous rocks

Lithology	Sampling sections	Results (Ma)	Strata (Formation)	Rock types	Methods	References
Intrusive units	Bachu	284.3±2.8	Syenite porphyry dike	Syenite porphyry	Zircon SHRIMP	Yang et al. (2007)
	Bachu	273.7±1.5	Intrusive complex	Syenite	Zircon SHRIMP	Zhang et al. (2008)
	Bachu	281±4	Syenite body	Syenite	Zircon LA-ICP-MS	Li et al. (2007)
	Bachu	277.7±1.3	Syenite body	Syenite	$^{40}Ar/^{39}Ar$	Yang et al. (1996)
	Bachu	277±4	Syenite body	Syenite	Zircon SHRIMP	Yang et al. (2006)
	Bachu	282±3	Syenite body	Syenite	Zircon LA-ICP-MS	Li et al. (2007)
	Bachu	285.9±2.6	Syenite body	Syenite	Zircon SHRIMP	Sun et al. (2008)
	Bachu	281.7±4.8	Syenite body	Syenite	Zircon LA-ICP-MS	Li et al. (2009)
	Bachu	279.7±2.0	Syenite body	Syenite	Zircon SIMS	Wei and Xu (2011)
	Bachu	283.1±3.2	Intrusive complex	Gabbro	Zircon LA-ICP-MS	Zhang et al. (2009)
	Bachu	278.9±2.1	Intrusive body	Wehrlite	Zircon SIMS	Wei et al. (2014b)
	Bachu	275.2±1.2	Intrusive complex	Diorite	Zircon SHRIMP	Zou et al. (2015)
	Bachu	272±6	Dike	Diabase	Zircon LA-ICP-MS	Li et al. (2007)
	Bachu	278.4±2.1	Intrusive body	Wehrlite	Zircon SIMS	Wei et al. (2014b)
	Bachu	281.4±1.7	Intrusive complex	Alkali diabase	Zircon SHRIMP	Zou et al. (2015)
	Kuche river	278±23	Xiaotiekanlike	Rhyolite	$^{40}Ar/^{39}Ar$	Yang et al. (1996)

(To be continued)

(Table 2.1)

Lithology	Sampling sections	Results (Ma)	Strata (Formation)	Rock types	Methods	References
Upper P$_{kpz}$ basalt	Keping	274.1±2.4	Kaipaizilaike Fm.	Basalt	^{40}Ar/^{39}Ar	Zhang et al. (2010)
	Keping	271.9±3.7	Kaipaizilaike Fm.	Basalt	^{40}Ar/^{39}Ar	Zhang et al. (2010)
	Yingan	287.3±4.0	Kaipaizilaike Fm.	Basalt	^{40}Ar/^{39}Ar	Wei et al. (2014a)
	Yingan	287.9±4.1	Kaipaizilaike Fm.	Basalt	^{40}Ar/^{39}Ar	Wei et al. (2014a)
	Yingan	287.2±5.6	Kaipaizilaike Fm.	Basalt	^{40}K/^{39}Ar	Yang et al. (2006)
	Taxinan	284±4	Qipan Fm. (interlayer)	Sandstone	Zircon LA-ICP-MS	Li et al. (2013)
	Taxinan	289.6±5.6	Qipan Fm.	Basalt	^{40}Ar/^{39}Ar	Yang et al. (2006)
	YM8	290.5±4.2	5376 m	Basalt	^{40}K/^{39}Ar	Huang et al. (2003)
Lower P$_{kpk}$ basalt	YM5	287	5420–5480 m (bottom interlayer)	Rhyolite	Zircon LA-ICP-MS	Tian et al. (2010)
	YM30	291	5410–6420 m (bottom interlayer)	Rhyolite	Zircon LA-ICP-MS	Tian et al. (2010)
	HLHT2	287.2±2.0	5430–5460 m (top interlayer)	Trachydacite	zircon SIMS	Shangguan et al. (2015)
	Yingan	289.0±6.1	Kupukuziman Fm. (Layer 6)	Basalt	^{40}K/^{39}Ar	Zhang et al. (2003)
	Yingan	281.8±4.2	Kupukuziman Fm.	Basalt	^{40}Ar/^{39}Ar	Yang et al. (2006)
	Sishichang	278.5±1.4	Kupukuziman Fm.	Basalt	^{40}Ar/^{39}Ar	Yang et al. (1996)

2.3.1 Basalts

Because the data for the previous age obtained for the Permian basalts and igneous rocks were poor, $^{40}Ar/^{39}Ar$ dating on basalts, and zircon SHRIMP and LA-ICP U–Pb dating were carried out on breccia, gabbro, diabase, diorite, rhyolite, and syenite from the northwestern and southwestern parts of the Tarim Basin.

The fact that the eruption times of the basalt from the Yingan section in the northwestern part of the Tarim Basin and from the Damusi section of the southwestern part of the Tarim Basin are synchronous can be deduced based on the recent $^{40}Ar/^{39}Ar$ dating. Through the $^{40}Ar/^{39}Ar$ dating from the third basaltic layer of the 4 basaltic layers of the Yingan section, and from the interlayer of the Damusi section on whole-rock basalts from the Yingan section, we yielded a plateau age of (281.8±4.2) Ma (2σ) and anti-isochron age of (282.5±4.4) Ma (2σ) for the Yingan basalt, and a weighted mean value of (290.1±3.5) Ma for the Damusi basalt (Fig. 2 of Yang et al., 2006a). The data regarding ages from the Kupukuziman and Kaipaizileike basalts were concentrated at 288–258 Ma with a mean value of 272 Ma, and the Kupukuziman basalts were concentrated in 292–278 Ma with a mean value of 285 Ma. Furthermore, a K–Ar age of (289.6±5.6) Ma (Li et al., 2008) and an $^{40}Ar/^{39}Ar$ weighted mean age of (290.1±3.5) Ma (Yang et al., 2006a) have been reported for Permian basalts exposed in the Qipan Section in the southwestern Tarim area.

Samples of YG20-21 and Txn25-21 for the $^{40}Ar/^{39}Ar$ age dating were from the third basaltic layer among a total of 4 basaltic layers of the Yingan section, and from the interlayer of the Damusi section (Yang et al., 2006a, b). The 2 samples are mainly made up of a groundmass of microlites of plagioclase, clinopyroxene and iron oxides with small amounts of olivine. The groundmass has intersertal and crytocrystalline textures, and massive structures. Both of the 2 samples had no phenocrysts, vesicular and amygdaloidal structures that can be used to measure $^{40}Ar/^{39}Ar$ ages (Yang et al., 2006a).

$^{40}Ar/^{39}Ar$ dating conducted on whole-rock basalts from the Yingan section yielded a plateau age of (281.8±4.2) Ma (2σ) including 35.3% of the ^{39}Ar and an anti-isochron age of (282.5±4.4) Ma (2σ) with MSWD = 0.26 for sample YG20-21, and a weighted mean value of (290.1±3.5) Ma for sample Txn25-21 in the Ar/Ar age diagrams (Yang et al., 2006a). $^{40}Ar/^{39}Ar$ stepwise heating data of whole-rocks from the Tarim basalt samples can be seen in Table 2 of Yang et al. (2006a).

Precise $^{40}Ar/^{39}Ar$ dating of these rocks revealed an eruption age span of 285–262 Ma (Zhang Y.-T. et al., 2010). Among them, 5 samples have ages of 285–283 Ma, and 3 samples have ages ranging of 274–271 Ma. Although the age span of these basaltic rocks seems large, Zhang Y.-T. et al. (2010) assumed that it is within the age distribution obtained in other works (Chen et al., 1997, 1998; Jia, 1997; Yang et al., 2006a, 2007a; Li, 2007).

Chen et al. (1997) argued that basalts, diabases and other intrusive rocks are products of Permian rifting based on their chronological ages (292±0.5) Ma and (259.8±0.9) Ma, field occurrence, geochemistry and Late Paleozoic sedimentary features in the interior of the Tarim Basin and the surrounding tectonic framework. Zhang et al. (2003) argued that the Kupukuziman and the Kaipaizileike formations were formed during the Lower–Middle and Middle Permian periods based on some evidence of paleontology and sediments. Some chronological studies ($^{40}Ar/^{39}Ar$, $^{40}K/^{39}Ar$, Sm–Nd) of borehole basalt samples from the central part of the Tarim Basin and other Permian igneous rocks as well as their geochemical characteristics indicate that the timing of this thermal overprinting on the basalts is related to the widespread Early Permian thermo–tectonic event in different parts of the Tarim Basin under within-plate environment, and they are regarded as occurring much earlier than the age data from previous studies.

The new $^{40}Ar/^{39}Ar$ age data gave clearer and more detailed information of the temporal variation of basaltic volcanism than the preliminary $^{40}K/^{39}Ar$ results of 292–248 Ma. In addition, the $^{40}Ar/^{39}Ar$ age data combined with previous isotopic data suggest that the most intensive and extensive tectono–thermal event occurred in the Early Permian age and no subsequent largely thermal disturbance (stable crustal setting) occurring in the interior of the Tarim Basin was clarified.

The above ages, as well as the ages that other researchers reported from the Tarim flood basalts, addressed a large tectono-magmatic event in the Tarim Basin that occurred during the Early Permian of 290–278 Ma.

Although the 2 basaltic sequences (Kai and Ku) are separated by a sedimentary succession of about 800 m thick (Fig. 1.7), zircon SHRIMP U–Pb dating of the basalts by Yu et al. (2011) gave the ages of (289.5±2.0) Ma in the 1st basaltic unit of the Kupukuziman Formation (Ku1) and (288.0±2.0) Ma in the 6th basaltic unit of the Kaipaizileike Formation (Kai6), indicating the ages of 290–288 Ma may represent the main basaltic eruption time.

2.3.2 Layered Intrusive Rocks

A whole-rock Sm–Nd isochron age of (306±7) Ma was reported for the Wajilitag intrusion (Lu et al., 2000). But Zhang et al. (2008) considered that the emplacement age of the intrusion might be ca. 275 Ma based on the zircon U–Pb dating of the syenite in the Bachu area. In addition, some olivine-rich ultramafic dikes with the trend of the north and northwest direction occurred in the Xiaohaizi area (known also as the Mazhaertage or Mazhartag area), north of the Wajilitag intrusion (Jiang et al., 2004c; Yang et al., 2007b; Zhou et al., 2009). According to the available geochronological data, Zhang C.-L. et al. (2010) suggested that these mafic–ultramafic intrusion and dikes, along with other igneous rocks in the Wajilitag region, were probably the product of a Permian upwelling mantle plume under the Tarim Block.

Zhang et al. (2008) have done U–Pb zircon dating on 20 zircon grains with U concentrations ranged from 68 to 225 ppm, Th concentrations from 36 to 115 ppm, and Th/U ratios from 0.42 to 0.62. In the $^{206}Pb/^{238}U$–$^{207}Pb/^{235}U$ concordia plot, the remaining 16 zircon analyses are concordant within errors (Fig. 3 of Zhang et al., 2008), yielding a weighted mean $^{206}Pb/^{238}U$ age of (274±2) Ma (MSWD=2.4). This age is interpreted as the timing of the BLIC (Bachu layered intrusive complex) emplacement. It is noted that this emplacement age is identical with a zircon SHRIMP U–Pb age of (272±1.2) Ma for a quartz syenite dike and a whole-rock Sm–Nd isochron age of (259±57) Ma for a mafic dike south of the layered intrusive complex (Yang et al., 2006), and this is close to a $^{40}Ar/^{39}Ar$ plateau age of (278.5±1.4) Ma for the Permian basalts in central Tarim (Chen et al., 1997; Jia, 1997).

2.3.3 Breccias

The new geochronological data from the Wajilitag kimberlitic intrusions supported the theory of the arrival of the mantle plume beneath the Tarim lithosphere at least 10 million years before the onset of the Tarim flood basalt volcanism (Zhang et al., 2013). Several tens of kimberlitic pipes and dikes are exposed in the Wajilitag area in the western Tarim Basin. Preliminary studies have suggested that the Wajilitag kimberlitic intrusions formed during the Late Permian (ca. 253 Ma) based on $^{40}Ar/^{39}Ar$ phlogopite dating (Li et al., 2001), and it may indicate a mixed age or cooling/resetting age. Zhang et al. (2013) recently obtained new spectrometric U–Pb age data on perovskite and baddeleyite grains in a kimberlitic pipe and a kimberlitic dike from the Tarim Block by the Secondary Ion Mass. The perovskite yielded a well-defined intercept age of (299.8±4.3) Ma, which is consistent with its corresponding concordia ^{206}Pb/^{238}U ages of about 300 Ma. The baddeleyite separated from two kimberlitic samples from a dike displays identical concordia U–Pb ages of (300.8±4.7) Ma and (300.5±4.4) Ma. These new ages are slightly older than the ages of the eruption of the Tarim flood basalts (291–274 Ma; Li et al., 2011), offering a critical regional time marker for the onset of the Permo–Carboniferous magmatism in the Tarim Block. Detailed petrographic observations did not reveal any ultrahigh-pressure mineral assemblage in the Wajilitag kimberlitic intrusions; and their new geochronological data suggested the arrival of the mantle plume beneath the Tarim lithosphere at least 10 million years before the onset of the Tarim flood basalt volcanism. The end-Carboniferous Wajilitag kimberlitic intrusions, the oldest known phase associated with Carboniferous magmatism in the Tarim Craton, signals the initial magmatic pulse triggered by mantle plume impingement.

The perovskites show relatively uniform uranium contents of (112±29) ppm and a Th/U ratio from 8.0 to 35.5. The scattered data points on the Tera-Wasserburg plot gave a well-defined lower intercept age at (299.8±7.5) Ma and an upper intercept with ^{207}Pb/^{206}Pb of 0.85±0.03 for the common-Pb composition; therefore, the corrected data yield a concordia U–Pb age of (299.8±4.3) Ma. ^{206}Pb/^{238}U individual dates, followed by the ^{207}Pb-based common-Pb correction, yield a weighted average ^{206}Pb/^{238}U age of (299.2±4.3) Ma (MSWD=0.62). Baddeleyite grains analyzed are mostly subhedral or fragmental, ranging from 40 to 100 mm in length (Fig. 4 of Zhang et al., 2013). The results show variable U contents from 59 to 790 ppm and Th/U from 0.02 to 0.20, with the exception of one analysis which yields a relatively high Th/U=0.37 due to a significantly high Th content (221 ppm) compared to other studied baddeleyite grains. The data yield identical concordia U–Pb ages of (300.8±4.7) Ma (MSWD=1.6) and (300.5±4.4) Ma (MSWD=1.3) (Zhang et al., 2013).

2.3.4 Picrite

As introduced in subsection 2.3.7, the rhyolites can be subdivided into two coeval groups with overlapping U–Pb zircon ages between (291±4) and (272±2) Ma, indicating the formation of picrite. The picrite–basalt–rhyolite suite from the NTU, together with Permian volcanic rocks from elsewhere in the Tarim Basin, constitutes an LIP that is characterized by a large areal extent, rapid eruption, an OIB-type chemical composition, and an eruption of high-temperature picritic magma (Tian et al., 2010).

2.3.5 Syenite Bodies

The quartz syenitic porphyritic dike occurred in Shuigongtuan of Bachu County, Tarim Basin. The concordia age of (277±4) Ma was obtained from the Xiaohaizi syenitic body by the SHRIMP zircon U–Pb age dating system (Yang et al., 2006), indicating that the intrusive age of the Xiaohaizi syenitic body is approximately 277 Ma. The ages for the layered intrusive rocks and syenitic rocks of 284–275 Ma were obtained by whole-rock $^{40}Ar/^{39}Ar$ and SHRIMP zircon U–Pb dating (Yang et al., 1996; 2007a; Zhang et al., 2008; Li et al., 2011), and are regarded to have formed during the Early Permian.

2.3.6 Quartz Syenite Porphyry

A fine-grained diked quartz syenite porphyry sample (sgt0503-1a) from the Xiaohaizi area was used for zircon U–Pb dating by the SHRIMP method. Zircons were separated from the sample using conventional crushing and mineral separation procedures. Zonal structures of selected zircon grains were documented using Cathodoluminescence (CL) images. The SHRIMP U–Pb isotopic measurements were carried out in the Beijing SHRIMP II Center, at the Chinese Academy of Geological Sciences.

Zircons from our sample (sgt0503-1a) are euhedral and ~250 μm in length. They can be divided into two types: (1) a brownish color, fractured, weak transparent and a length/width ratio between 3:1 and 2:1; (2) a pale yellow color, clearly transparent, fine-grained, a length/width ratio of ~3:1, with small magnetite inclusions in some grains. A total of 150 grains of both types of zircons with polygonal zoning were selected for U–Pb dating.

A total of 21 spot analyses were obtained from 21 selected zircon crystals. The zircon grains contain 112–264 ppm U and 67–206 ppm Th. Th/U ratios of the zircon grains are from 1.3 to 2.3, significantly higher than the value (>0.22) of typical magmatic zircon. Among a total of 21 analyses, one analysis yields a $^{206}Pb/^{238}U$ age of 273 Ma (^{204}Pb corrected), and the other 20 analyses collectively yield a weighted mean $^{206}Pb/^{238}U$ age of (284±2.8) Ma (MSWD=1.38).

The new zircon U–Pb age of (284±2.8) Ma for the Xiaohaizi quartz syenite porphyry is ~10 Ma older than the ages of ~274 Ma for the Wajilitag quartz syenite and associated mafic dikes reported previously (Li et al., 2007; Zhang et al., 2008). Previous studies indicate that the syenite and mafic dikes in the Xiaohaizi and Wajilitag areas have similar ages between 278 and 273 Ma (Yang et al., 1997, 2006b; Li et al., 2007; Zhang et al., 2008). Our new result extends the lower limit of the age range for the syenite-diabase association to 284 Ma. Therefore, the age of the syenite-diabase association ranges from 284 to 274 Ma (Li et al., 2011).

Based on the present study, the Shuigongtuan quartz syenitic porphyritic dike and Xiaohaizi syenitic body probably formed during the Early Permian (277 Ma), and represented the products of within-plate environments, indicating that the end of the last large magmatic thermal event occured in the Tarim Basin (Yang et al., 2006a).

The timing of intrusion of the dikes is an important aspect to be resolved. Yang et al. (2007a) proposed that the quartz syenite porphyry in the bimodal dike may have an intrusion age of ca. 277 Ma similar to that of the surrounding syenite body ($^{40}Ar/^{39}Ar$ age of (277.7±1.3) Ma quoted from Yang et al., 1996) and diabase dikes. Therefore, the quartz syenite porphyry dike in the bimodal suite is inferred to have formed in the same age of the diabase dike (Yang et al., 2007).

A combination of data from the present study together with those of Yang et al. (1997) and Chen et al. (1997), suggests that the Tarim Basin evidenced a major tectonothermal event in the Permian, with strong imprints in the central and western parts of the Basin. The bimodal dike, Xiaohaizi syenite, basalt, diabase and syenite constitute a suite of magmatic rocks that formed in a rifting environment implying further that a large continental rifting setting developed in the interior of the Tarim basin during the Permian (Yang et al., 2007).

The quartz syenite porphyry and diabase dikes constitute the bimodal dike in Bachu Shuigongtuan, with a sharp direct contact and bimodal association. The dikes are inferred to have formed under a rift environment at ca. 277 Ma or slightly later, and probably represent a major magmatic event in the Tarim Basin (Yang et al., 2007).

2.3.7 Rhyolite in NTU

Tian et al. (2010) did the LA-ICP-MS zircon U–Pb dating for rhyolites in the NTU as follows. Rhyolites interlayered with basalts are interpreted to be roughly coeval with their surrounding flood basalts. Two such samples of rhyolites were targeted in an attempt to date the lifespan of basalt magmatism. They are YM30-1, from a depth of 6330 m, which is the lowest intersected rhyolite, and YM5-8 from 5484 m, near the top of the basalt sequence. These two samples enclose a stratigraphic basalt thickness of over 1200 m in the borehole YM30, and over 2500 m in the boreholes YM8 and YM5, as clearly indicated from seismological cross sections. YM30-1 yielded zircons with euhedral and prismatic shapes, sharp terminations, and a concentric oscillatory zoning of magmatic origin. Interpretation of the age of this rhyolite is relatively straightforward. The remaining population yielded a weighted mean age of (290.9±4.1) Ma (MSWD=1.62) for 18 out of 40 analyzed zircons. YM5-8, the sample close to the top of the pile, yielded a weighted mean age of (286.6±3.3) Ma for 16 out of 38 analyzed zircons (MSWD=1.04).

2.3.8 Temporal Relations of Different Rock Units in the Tarim LIP

Based on field relationships and the results of the $^{40}K/^{39}Ar$, Rb–Sr and Sm–Nd radiometric dating, the previous researchers (e.g., Zhang et al., 2003; Jia, 1997; Yang et al., 1996a) suggested that the basalts and diabases in the Tarim Basin formed between 300 Ma and 260 Ma. Zhang et al. (2003) reported K–Ar ages of 285–260 Ma for basalts in the Keping area. Yang et al. (2006a) reported whole Ar–Ar isotopic ages of (281.8±4.2) Ma, (278.5±1.4) Ma and (289.6±5.6) Ma for the basalts in the Yingan, Sishichang (Ssc) and Taxinan (Txn) areas, respectively.

We propose that the Kupukuziman and Kaipaizileike basalts were concentrated between 290–285 Ma and formed during the Early Permian. The precise age data and stratigraphic correlations could be used to evaluate the temporal evolution of the Tarim LIP. Although geochronological data for different rock units of the Tarim LIP have been reported (Liu et al., 1991; Jia et al., 1992; Yang et al., 1996, 2006a; Zhang et al., 2003; Li et al., 2007; Zhang et al., 2008; Yu et al., 2011), the age results using different dating methods and some data obtained from the 1990s as well as the difficulty in observing the field contact relationship made it so that the sequence of magmatism is still unclear. The basalts of the Tarim LIP in the Yingan section of the Keping area and the Damusi section have $^{40}Ar/^{39}Ar$ ages of 282 Ma and 290–288 Ma (Yang et al., 2006a). Whole-rock basaltic samples from the third lava flows in the Yingan section, and from the Damusi section yielded an $^{40}Ar/^{39}Ar$ plateau age of (281.8±4.2) Ma (2σ) with an anti-isochron age of (282.5±4.4) Ma (2σ), and a weighted mean value of (290.1±3.5) Ma, respectively (Yang et al., 2006a). The $^{40}K/^{39}Ar$ ages of the basaltic samples from the Yingan section (Keping area) and the Damusi section (Taxinan area) are between (287.2±5.6) Ma (unpublished) and (289.6±5.6) Ma (Li et al., 2008), consistent with the $^{40}Ar/^{39}Ar$ dating results, in which the samples analyzed are the same. In general, the zircon U–Pb and $^{40}Ar/^{39}Ar$ ages of the basaltic rocks support the theory that the formation of the basalts in the Kupukuziman, Kaipaizileike and Qipan formations was concentrated in the period of 290–285 Ma. Yu et al. (2011) obtained the ages of (289.5±2.0) Ma and (288.0±2.0) Ma by SHRIMP zircon U–Pb dating from the basalts from the bottom of the Kupukuziman Formation and the top of the Kaipaizileike Formation in the Yingan section in the Keping area. These ages indicate the onset and the main period of the eruptions of the basalts though there is some debate as to the explanation of zircon ages as the formation age of the basalts.

Using the LA-ICP-MS method, Li et al. (2007) obtained zircon U–Pb ages of (282±3) Ma and (281±4) Ma for the Wajilitag quartz syenite and (272±6) Ma for diabase in the Bachu area. These data suggest that the syenite was emplaced slightly before the mafic dike in the region. Most recently, Zhang et al. (2008) reported a SHRIMP zircon U–Pb age of (274±2) Ma for the Wajilitag quartz syenite.

Newly obtained age data and field evidence suggest that the intrusive rocks, including the layered mafic–ultramafic intrusive bodies, breccia mica-olivine pyroxenite pipes, diabase dike swarms, syenite intrusions and quartz syenite porphyry formed between 284 and 274 Ma after the eruption of the basalt lavas (Li et al., 2011). Zhang et al. (2009) gave a LA-ICP-MS zircon U–Pb age of (283.1±3.2) Ma and (265±16) Ma from gabbro samples of the layered mafic–ultramafic intrusive bodies, and the latter has a big error, indicating that the age of 283 Ma is much more reliable for the formation of the gabbro. Zhang et al. (2009) and Li et al. (2005) reported a LA-ICP-MS zircon U–Pb age of (283±1.3) Ma and (272±6) Ma for a diabase in the Keping area, respectively; and these data coincide with the field evidence showing that the diabases formed after the eruption of the basalt lavas (Chen et al., 2006). Zhang et al. (2008) reported a SHRIMP zircon U–Pb age of 274 Ma for a quartz syenite body in the Wajilitag area, which is ~4 Ma younger than the SHRIMP zircon age and whole-rock $^{40}Ar/^{39}Ar$ age given by Yang et al. (2007a) and Yang et al. (1996). However, the age result of (281.7±4.8) for a quartz syenite body (Zhang et al., 2009) did not support the age of 274 Ma (Zhang et al., 2008). We argue that the formation of quartz syenite should range from

284–274 Ma. Yang et al. (1996) reported a $^{40}Ar/^{39}Ar$ age of (277.7±1.3) Ma for the syenite, being within the range of the age of 284–274 Ma. The new data show a much older age of 284 Ma for the quartz syenite porphyry than those from the quartz syenite body (Yang et al., 2006b; Zhang et al., 2008; Zhang et al., 2009), but the field evidence shows that the quartz syenite body was much older than the quartz syenite porphyry. SHRIMP zircon U–Pb dating indicates that the Xiaohaizi quartz syenite and quartz syenite porphyry have similar ages of 284–274 Ma. Therefore, we argue that the ages of the quartz syenite and the quartz syenite porphyry concentrate in 284–274 Ma and have a much wide range for the time of their formation.

Zhang C.-L. et al. (2010) obtained Zircon U–Pb ages of ca. 276 Ma and ca. 278 Ma from the Piqiang oxide-bearing ultramafic–mafic complex and the Halajun A-type granites in the Tarim Block respectively, indicating an emplaced time for their formation. Together with previously reported geochronological data, the diverse intrusive and extrusive rocks in the Tarim Basin show a peak age at ca. 275 Ma.

Permian plant fossils are present in mudstones in-bedded between basaltic lavas of the Kupukuziman and Kaipaizileike formations in the Shajingzi and Sishichang sections (Sun and Shen, 1991). Permian plant fossils also were found in the lower parts of the Kaipaizileike Formation in the Shajingzi section, which is equivalent to the lower Shihezi Formation of the Early Permian age in the North China craton (Zhang et al., 2003). Zhang et al. (2003) reported large amounts of paleontological fossils in the Kupukuziman and Kaipaizileike formations, and argued that the Kupukuziman Formation formed during the early to middle Early Permian and the Kaipaizileike Formation formed during the late Early Permian based on the occurrences of the *Whiphlella–Darwinula* assemblage, big plant fossils of *Autunia conferta–Pecopertis–Cordaites* and *Dichophyllum flabellifera* assemblages, and *Autunia conferta–Pecopertis–Cordaites*, *Dichophyllum flabellifera* and *Sphenophyllum verticillatum–"Noeggerathiopsis" subangusta* assemblages. In the Qipan Formation of the Taxinan area, there is a *Liraplecta aspera–Choristites tarimensis* assemblage and a *Potonieisporites–Vestigisporites* (PV) sporopollen assemblage indicating the age of early Early Permian, corresponding to the Qixia Formation in the North China Plate. Zhang et al. (2003) argued that the Kupukuziman and Kaipaizileike formations belong to the lower Permian based on the sedimentary sequence and fossil evidence.

Based on the new zircon SHRIMP U–Pb and recently published whole-rock $^{40}Ar/^{39}Ar$ and LA-ICP-MS zircon U–Pb age dating, the time sequence of magmatism of the Tarim LIP in the central and western part of the Tarim Basin is as follows (Fig. 2.4): basaltic lavas in the Kupukuziman Formation and then Kaipaizileike Formation (290–285 Ma), layered mafic–ultramafic intrusion, breccia mica-olivine pyroxenite pipes, diabase dike swarms, syenite intrusions and quartz syenite porphyry (284–274 Ma).

Through the study of the temporal relations of the different rock types in the Tarim LIP using stratigraphic correlation, lithologically spatial distribution and isotopic ages, and combined with the recent SHRIMP and LA-ICP-MS zircon U–Pb and $^{40}Ar/^{39}Ar$ age dating, we suggested the different rock units in the Tarim LIP formed between 290 and 274 Ma, and these data are more reliable than the K–Ar ages of 310–220 Ma reported previously. The sequence of magmatism of the Tarim LIP in the central and western parts of the Tarim Basin are basaltic lava in the Kupukuziman and Kaipaizileike Formations (290–285 Ma), layered mafic–ultramafic rock, mica-olivine pyroxenite breccia pipe, diabase and ultramafic dike, quartz syenite, quartz syenite porphyry and bimodal dike (284–274 Ma) (Li et al., 2011).

With the new data from the Tarim event, it is now possible to consider genetic links among the Tarim (280 Ma), Emeishan (258 Ma) and Siberian events (250 Ma) (R.E. Ernst, personal communication, 2006). The existence of three short magmatic pulses at 280, 258 and 250 Ma would seem to indicate three discrete events in the source region, and it is reasonable to infer that all three discrete LIP events originate as discrete plumes arising from the deep mantle. However, the broad spatial grouping of three major plume/LIP events within a 30-Ma period strongly suggests a genetic link between these, and it can be inferred that all three are related to a single protracted thermal anomaly (caused by heat loss from the core) in a region of the deep mantle (Yang et al., 2006b).

References

Ao, S.-J., Xiao, W.-J., Han, C.-M., Mao, Q.-G., Zhang, J.-E., 2010. Geochronology and geochemistry of Early Permian mafic–ultramafic complexes in the Beishan area, Xinjiang, NW China: Implications for late Paleozoic tectonic evolution of the southern Altaids. Gondwana Research, 18: 466-478.

Bao, P.-S., Su, L., Zhai, Q.-G., Xiao, X.-C., 2009. Compositions of the kimberlitic brecciated peridotite in the Bachu area, Xijiang and its ore-bearing potentialities. Acta Geol. Sin., 83: 1276-1301 (in Chinese with English abstract).

Blake, T.S., 2004. Geochronology of a Late Archean flood basalt province in the Pilbara Craton, Australia: Constraints on basin evolution, volcanic and sedimentary accumulation, and continental drift rates. Precambrian Research, 133: 143-173.

Bryan, S.E., Ernst, R.E., 2008. Revised definition of Large Igneous Provinces (LIPs). Earth-Science Reviews, 86(1-4): 175-202.

Chen, H.-L., Yang, S.-F., Dong, C.-W., et al., 1997. The discovery of Early Permian basic rock belt in Tarim Basin and its tectonic meaning. Geochemica, 26(6): 77-87 (in Chinese with English abstract).

Chen, H.-L., Yang, S.-F., Jia, C.-Z., Dong, C.-W., Wei, G.-Q., 1998. Confirmation of Permian intermediate-acid igneous rock zone and a new understanding of tectonic evolution in the northern part of the Tarim Basin. Acta Mineralogica Sinica, 18(3): 370-376 (in Chinese with English abstract).

Chen, H.-L., Yang, S.-F., Wang, Q.-H., Luo, J.-C., Jia, C.-Z., Wei, G.-Q., Li, Z.-L., He, G.-Y., Hu, A.-P., 2006. Sedimentary response to the Early–Mid Permian basaltic magmatism in the Tarim plate. Geology in China, 33(3): 545-552 (in Chinese with English abstract).

Chen, N.-H., Dong, J.-J., Yang, S.-F., Chen, J.-Y., Li, Z.-L., Ni, N.-N., 2014. Restoration of geometry and emplacement mode of the Permian mafic dike swarms in Keping and its adjacent areas of the Tarim Block, NW China. Lithos, 204(3): 73-82.

Chen, Z.-Q., Shi, G.-R., 2003. Late Paleozoic depositional history of the Tarim Basin, Northwest China: An integration of biostratigraphic and lithostratigraphic constraints. AAPG Bulletin, 87(8): 1323-1354.

Chung, S.L., Jahn, B.M., 1995. Plume–lithosphere interaction in generation of the Emeishan flood basalts at the Permian–Triassic boundary. Geology, 23: 889-892.

Coffin, M.F., Eldholm, O., 1994. Large Igneous Provinces: Crustal structure, dimensions, and external consequences. Rev. Geophys., 32: 1-36.

Dobretsov, N.L., 2005. 250 Ma Large Igneous Provinces of Asia: Siberian and Emeishan traps (plateau basalts) and associated granitoids. Russian Geology and Geophysics, 46(9): 847-868.

Donnelly, T.W., Rogers, J.J.W., 1980. Igneous series in island arc: The northeastern Caribbean compared with worldwide island-arc assemblages. Bulletin Volcanologique, 43: 347-382.

Fekiacova, Z., Abouchami, W., Galer, S.J.G., Garcia, M.O., Hofmann, A.W., 2007. Origin and temporal evolution of Ko'olau Volcano, Hawaii: Inferences from isotope data on the Ko'olau Scientific Drilling Project (KSDP), the Honolulu Volcanics and ODP Site 843. Earth and Planetary Science Letters, 261(1-2): 65-83.

Geist, D., Howard, A., Larson, P., 1995. The generation of oceanic rhyolites by crystal fractionation: The basalt-rhyolite association at Volcan Alcedo, Galapagos Archipelago. Journal of Petrology, 36: 965-982.

Guo, Z.-J., Yin, A., Robinson, A., Jia, C.-Z., 2005. Geochronology and geochemistry of deep-drill-core samples from the basement of the central Basin. Journal of Asian Earth Sciences, 25(1): 45-56.

Halls, H.C., Davis, D.W., Stott, G.M., Ernst, R.E., Hamilton, M.A., 2008. The Paleoproterozoic Marathon Large Igneous Province: New evidence for a 2.1 Ga long-lived mantle plume event along the southern margin of the North American Superior Province. Precambrian Research, 162: 327-353.

He, B., Xu, Y.-G., Huang, X.-L., Luo, Z.-Y., Shi, Y.-R., Yang, Q.-J., Yu, S.-Y., 2007. Age and duration of the Emeishan flood volcanism, SW China: Geochemistry and SHRIMP zircon U–Pb dating of silicic ignimbrites, post-volcanic Xuanwei Formation and clay tuff at the Chaotian section. Earth and Planetary Science Letters, 255(3-4): 306-323.

Hochstaedier, A.G., Gill, J.B., Morris, J., 1990. Volcanism in the Sumisu Rift, II. Subduction and non-subduction related components. Earth and Planetary Science Letters, 100: 195-209.

Hu, A.-Q., Jahn, B.M., Zhang, G.-X., Zhang, Q.-F., Chen, Y.-B., 2000. Crustal evolution and Phanerozoic crustal growth in northern Xinjiang: Nd–Sr isotopic evidence. Part I: Isotopic characterisation of basement rocks. Tectonophysics, 328: 15-51.

Huang, H., Zhang, Z.-C., Kusky, T., Santosh, M., Zhang, S., Zhang, D.-Y., Liu, J.-L., Zhao, Z.-D., 2012. Continental vertical growth in the transitional zone between South Tianshan and Tarim, western Xinjiang, NW China: Insight from the Permian Halajun A1-type granitic magmatism. Lithos, 155: 49-66.

Jia, C.-Z., 1997. Tectonic Characteristics and Oil-Gas, Tarim Basin, China. Beijing: Petroleum Industry Press, p. 438 (in Chinese).

Jia, C.-Z., Yao, H.-Q., Gao, J., Zhou, D.-Y., Wei, G.-Q., 1992. The stratum system of Tarim Basin. In: Tong, X.-G., Liang, D.-G. (Eds.), The Article Volume of Oil and Gas Exploration of the Tarim Basin. Urumqi: Xinjiang Scientific, Technology and Sanitation Press (in Chinese).

Jia, C.-Z., Wang, L.-S., Wei, G.-Q, et al., 2004. Plate Tectonics and Continental Geodynamics of The Tarim Basin. Beijing: Petroleum Industry Press, pp. 127-135 (in Chinese).

Jiang, C.-Y., Zhang, P.-B., Lu, D.-R., Bai, K.-Y., 2004a. Petrogenesis and magma source of the ultramafic rocks at Wajilitag region, western Tarim Plate in Xinjiang. Acta Petrologica Sinica, 20: 1433-1444 (in Chinese with English abstract).

Jiang, C.-Y., Zhang, P.-B., Lu, D.-R., Bai, K.-Y., Wang, Y.-P., Tang, S.-H., Wang, J.-H., Yang, C., 2004b. Petrology, geochemistry and petrogenesis of the Kalpin basalts and their Nd, Sr and Pb isotopic compositions. Geology Review, 50: 492-500 (in Chinese with English abstract).

Jiang, C.-Y., Jia, C.-Z., Li, L.-C., Zhang, P.-B., Lu, D.-R., Bai, K.-Y., 2004c. Source of the Fe-riched-type high-Mg magma in Mazhartag region, Xinjiang. Acta Geologica Sinica, 78(6): 770-780 (in Chinese with English abstract).

Jiang, C.-Y., Li, Y.-Z., Zhang, P.-B., Ye, S.-F., 2006. Petrogenesis of Permian basalts on the western margin of the Tarim Basin, China. Russian Geology and Geophysics, 47: 237-248.

Jordan, B.T., Grunder, A.L., Duncan, R.A., Deino, A.L. 2004. Geochronology of age-progressive volcanism of the Oregon High Lava Plains: Implications for the plume interpretation of Yellowstone. Journal of Geophysical Research-Solid Earth, 109: B10202.

Kryza, R., Zalasiewicz, J., 2008. Records of Precambrian–Early Palaeozoic volcanic and sedimentary processes in the Central European Variscides: A review of SHRIMP zircon data from the Kaczawa succession (Sudetes, SW Poland). Tectonophysics, 461(1-4): 60-71.

Li, C.-N., Lu, F.-X., Chen, M.-H., 2001. Research on petrology of the Wajilitag complex body in north edge in the Tarim Basin. Xinjiang Geology, 19: 38-43 (in Chinese with English abstract).

Li, W.-X., Li, X.-H., Li, Z.-X. 2005. Neoproterozoic bimodal magmatism in the Cathaysia Block of South China and its tectonic significance. Precambrian Research, 136: 51-66.

Li, X.-H., Li, Z.-X., Zhou, H., Liu, Y., Kinny, P.D., 2002. U–Pb zircon geochronology, geochemistry and Nd isotopic study of Neoproterozoic bimodal volcanic rocks in the Kangdian Rift of South China: Implications for the initial rifting of Rodinia. Precambrian Research, 113: 135-154.

Li, Y., Su, W., Kong, P., Qian, Y.-X., Zhang, K.-L., Zhang, M.-L., Chen, Y., Cai, X.-Y., You, D.-H., 2007. Zircon U–Pb ages of the Early Permian magmatic rocks in the Tazhong–Bachu region, Tarim Basin by LA-ICP-MS. Acta Petrologica Sinica, 23(5): 1097-1107 (in Chinese with English abstract).

Li, Y.-Q., Li, Z.-L., Sun, Y.-L., Chen, H.-L., Yang, S.-F., Yu, X., 2010. PGE and geochemistry of Wajilitag ultramafic cryptoexplosive brecciated rocks from Tarim Basin: Implications for petrogenesis. Acta Petrologica Sinica, 26(11): 3307-3318 (in Chinese with English abstract).

Li, Y.-Q., Li, Z.-L., Sun, Y.-L., Santosh, M., Langmuir, C.H., Chen, H.-L., Yang, S.-F., Chen, Z.-X., Yu, X., 2012. Platinum-group elements and geochemical characteristics of the Permian continental flood basalts in the Tarim Basin, Northwest China: Implications for the evolution of the Tarim Large Igneous Province. Chemical Geology, 328: 278-289.

Li, Y.-Q., Li, Z.-L., Yu, X., Langmuir, C.H., Santosh, M., Yang, S.-F., Chen, H.-L., Tang, Z.-L., Song, B., Zou, S.-Y., 2014. Origin of the early Permian zircons in keping basalts and magma evolution of the Tarim Large Igneous Province (northwestern China). Lithos, 204(3): 47-58.

Li, Z.-L., Yang, S.-F., Chen, H.-L., Langmuir, C.H., Yu, X., Lin, X.-B., Li, Y.-Q., 2008. Chronology and geochemistry of Taxinan basalts from the Tarim Basin: Evidence for Permian plume magmatism. Acta Petrologica Sinica, 24: 959-970 (in Chinese with English abstract).

Li, Z.-L., Chen, H.-L., Song, B., Li, Y.-Q., Yang, S.-F., Yu, X., 2011. Temporal evolution of the Permian Large Igneous Province in Tarim Basin in northwestern China. Journal of Asian

Earth Sciences, 42(5): 917-927.

Li, Z.-L., Li, Y.-Q., Chen, H.-L., Santosh, M., Yang, S.-F., Xu, Y.-G., Langmuir, C.H., Chen, Z.-X., Yu, X., Zou, S.-Y., 2012. Hf isotopic characteristics of the Tarim Permian Large Igneous Province rocks of NW China: Implication for the magmatic source and evolution. Journal of Asian Earth Sciences, 49: 191-202.

Li, Z.-X., Li, X.-H., Kinny, P.D., Wang, J., Zhang, S., Zhou, H., 2003. Geochronology of Neoproterozoic syn-rift magmatism in the Yangtze Craton, South China and correlations with other continents: Evidence for a mantle superplume that broke up Rodinia. Precambrian Res., 122: 85-109.

Lindline, J., Crawford, W.A., Crawford, M.L., 2004. A bimodal volcanic-plutonic system: The Zarembo Island extrusive suite and the Burnett Inlet intrusive complex. Canadian Journal of Earth Sciences, 41: 355-375.

Liu, H.-Q., Xu, Y.-G., Tian, W., Zhong, Y.-T., Mundil, R., Li, X.-H., Yang, Y.-H., Luo, Z.-Y., Shangguan, S.-M., 2014. Origin of two types of rhyolites in the Tarim Large Igneous Province: Consequences of incubation and melting of a mantle plume. Lithos, 204(3): 59-72.

Liu, J.-K., 1991. Petrological features and age of basalts from north part of Tarim Basin. In: Jia, Y.-X. (Ed.), Research of Petroleum and Gas Geology of Northern Tarim Basin in China. Wuhan: China University of Geoscience Press, pp. 194-201 (in Chinese).

Long, X.-P., Yuan, C., Sun, M., Kröner, A., Zhao, G.-C., Wilde, S., Hu, A.-Q., 2011. Reworking of the Tarim Craton by underplating of mantle plume-derived magmas: Evidence from Neoproterozoic granitoids in the Kuluketage area, NW China. Precambrian Research, 187(1-2): 1-14.

Long, X.-P., Yuan, C., Sun, M., Zhao, G.-C., Xiao, W.-J., Wang, Y.-J., Yang, Y.-H., Hu, A.-Q., 2010. Archean crustal evolution of the northern Tarim craton, NW China: Zircon U–Pb and Hf isotopic constraints. Precambrian Research, 180(3-4): 272-284.

Lu, F.-X., Li, C.-N., Chen, M.-H., 2000. Study on alkali rock belt and metallogenic geological conditions of rare earth elements, Gem and diamond in the north Tarim Basin. Xinjiang National 305 Project Office Open File, pp. 75-110 (in Chinese).

McNutt, S.R., Nishimura, T., 2008. Volcanic tremor during eruptions: temporal characteristics, scaling and constraints on conduit size and processes. Journal of Volcanology and Geothermal Research, 178(1): 10-18.

Morgan, J.P., Reston, T.J., Ranero, C.R., 2004. Contemporaneous mass extinctions, continental flood basalts, and "impact signals": Are mantle plume-induced lithospheric gas explosions the causal link? Earth and Planetary Science Letters, 217: 263-284.

Oberc-Dziedzic, T., Pin, C., Kryza, R. 2005. Early Paleozoic crustal melting in an extensional setting: petrological and Sm–Nd evidence from the Izera granite-gneisses, Polish Sudetes. International Journal of Earth Sciences, 94: 354-368.

Pang, K.-N., Zhou, M.-F., Lindsley, D.H., Zhao, D.-G, Malpas, J., 2008a. Origin of Fe–Ti oxide ores in mafic intrusions: Evidence from the Panzhihua intrusion. Journal of Petrology, 49:

295-313.

Pang, K.-N., Li, C.-S., Zhou, M.-F., Ripley, E.M., 2008b. Abundant Fe–Ti oxide inclusions in olivine from the Panzhihua and Hongge layered intrusions, SW China: Evidence for early saturation of Fe–Ti oxides in ferrobasaltic magma. Contributions to Mineralogy and Petrology, 156: 307-321.

Pang, K.-N., Li, C.-S., Zhou, M.-F., Ripley, E.M., 2009. Mineral compositional constraints on petrogenesis and oxide ore genesis of the Panzhihua layered gabbroic intrusion, SW China. Lithos, 110: 199-214.

Pirajno, F., Mao, J.-W., Zhang, Z.C., Zhang, Z.H., Chai, F.M., 2008. The association of mafic–ultramafic intrusions and A-type magmatism in the Tian Shan and Altai orogens, NW China: Implications for geodynamic evolution and potential for the discovery of new ore deposits. Journal of Asian Earth Sciences, 32: 165-183.

Pirajno, F., Ernst, R.E., Borisenko, A.S., Fedoseev, G., Naumov, E.A., 2009. Intraplate magmatism in Central Asia and China and associated metallogeny. Ore Geology Reviews, 35(2): 114-136.

Qian, Q., Wang, Y., 1999. Geochemical characteristics of bimodal volcanic suites from different tectonic settings. Geology and Geochemistry, 27: 29-32.

Qin, K.-Z., Su, B.-X., Sakyi, P.A., Tang, D.-M., Li, X.-H., Sun, H., Xiao, Q.-H., Liu, P.-P., 2011. SIMS zircon U–Pb geochronology and Sr–Nd isotopes of Ni–Cu-bearing mafic–ultramafic intrusions in Eastern Tianshan and Beishan in correlation with flood basalts in Tarim Basin (NW China): Constraints on a ca. 280 Ma mantle plume. American Journal of Science, 311(3): 237-260.

Shangguan, S.-M., Tian, W., Xu, Y.-G., Guan, P., Pan, L., 2012. The eruption characteristics of the Tarim flood basalt. Acta Petrologica Sinica, 28(4): 1261-1272 (in Chinese with English abstract).

Shao, J.-A., Li, X.-H., Zhang, L.-Q., Mou., B.-L., Liu, Y.-L., 2001. Geochemical condition for genetic mechanism of the Mesozoic bimodal dike swarms in Nankou-Guyaju. Geochemica, 30: 517-524.

Shu, L.-S., Zhu, W.-B., Wang, B., Faure, M., Charvet, J., Cluzel, D., 2005. The post-collision intracontinental rifting and olistostrome on the southern slope of Bogda Mountains, Xinjiang. Acta Petrologica Sinica, 21: 25-36.

Shu, L.-S., Deng, X.-L., Zhu, W.-B., Ma, D.-S., Xiao, W.-J., 2011. Precambrian tectonic evolution of the Tarim Block, NW China: New geochronological insights from the Quruqtagh domain. Journal of Asian Earth Sciences, 42(5): 774-790.

Su, L., 1991. A research on magmatic inclusions in minerals from kimberlites in Bachu County, Xinjiang, China. Bulletin of Xi'an Institute of Geology and Mineral Resources, Chinese Academy of Geological Sciences, 32: 33-46.

Sun, B.-N., Shen, G.-L., 1991. Discussion on Permian palaeophytogeographic province in northern margin of Tarim. In: Jia, Y.-X. (Ed.), Research of Petroleum and Gas Geology of Northern Tarim Basin in China. Wuhan: China University of Geoscience Press, pp. 186-

193 (in Chinese with English summary).

Sun, L.-H., Wang, Y.-J., Fan, W.-M., Zi, J.-W., 2008. A further discussion of the petrogenesis and tectonic implication of the Mazhashan syenites in the Bachu area. Journal of Jilin University, 38(1): 8-20.

Tian, W., Campbell, I.H., Allen, C.M., Guan, P., Pan, W.-Q., Chen, M.-M., Yu, H.-J., Zhu, W.-P., 2010. The Tarim picrite–basalt–rhyolite suite, a Permian flood basalt from northwest China with contrasting rhyolites produced by fractional crystallization and anatexis. Contributions to Mineralogy and Petrology, 160: 407-425.

Vorontsov, A.A., Yarmolyuk, V.V., Baikin, D.N., 2004. Structure and composition of the Early Mesozoic volcanic series of the Tsagan-Khurtei Graben, western Transbaikalia: Geological, geochemical, and isotopic data. Geochemistry International, 42: 1046-1061.

Wang, C.-Y., Zhou, M.-F., Zhao, D., 2008. Fe–Ti–Cr oxides from the Permian Xinjie mafic–ultramafic layered intrusion in the Emeishan Large Igneous Province, SW China: Crystallization from Fe- and Ti-rich basaltic magmas. Lithos, 102: 198-217.

Wang, Y.-S., Su, L., 1987. Petro-mineral characteristics of Wajilitag kimerlites and contrast with some correlation region, Bachu, Xinjiang. Periodical of Xi'an Geological Mineral Institue, 15: 47-56 (in Chinese).

Wang, Y.-S., Su, L. 1990. Composition characteristics and forming condition of phlogolites in Wajilitag komberlites, Bachu, Xinjiang. Periodical of Xi'an Geological Mineral Institute, 28: 47-55 (in Chinese).

Wei, X., Xu, Y.-G., 2011. Petrogenesis of Xiaohaizi syenite complex from Bachu area, Tarim. Acta Petrologica Sinica, 27: 2984-3004 (in Chinese with English abstract).

West, D.P., Coish, R.A., Tomascak, P.B. 2004. Tectonic setting and regional correlation of Ordovician metavolcanic rocks of the Casco Bay Group, Maine: Evidence from trace element and isotope geochemistry. Geological Magazine, 141: 125-140.

Wignall, P.B., 2001. Large Igneous Provinces and mass extinctions. Earth-Science Reviews, 53(1-2): 1-33.

Wilde, S.A., Wu, F.-Y., Zhang, X.-Z., 2003. Late Pan-African magmatism in northeastern China: SHRIMP U–Pb zircon evidence from granitoids in the Jiamusi Massif. Precambrian Research, 122(1-4): 311-327.

Xinjiang BGMR (Xinjiang Bureau of Geology and Mineral Resources), 1993. Regional geology of the Xinjiang Uygur Autonomous Region. Beijing: Geological Publishing House (in Chinese).

Xu, B., Jian, P., Zheng, H.-F., Zou, H.-B., Zhang, L.-F., Liu, D.-Y., 2005. U–Pb zircon geochronology and geochemistry of Neoproterozoic volcanic rocks in the Tarim Block of northwest China: Implications for the breakup of Rodinia supercontinent and Neoproterozoic glaciations. Precambrian Research, 136(2): 107-123.

Xu, B., Xiao, S.-H., Zou, H.-B., Chen, Y., Li, Z.-X., Song, B., Liu, D.-Y., Zhou, C.-M., Yuan, X.-L., 2009. SHRIMP zircon U–Pb age constraints on Neoproterozoic Quruqtagh diamictites in NW China. Precambrian Research, 168(3-4): 247-258.

Xu, Y.-G., Chung, S.-L., Jahn, B.M., Wu, G.-Y., 2001. Petrologic and geochemical constraints on the petrogenesis of Permian–Triassic Emeishan flood basalts in southwestern China. Lithos, 58: 145-168.

Xu, Y.-G., He, B., Huang, X.-L., et al., 2007. Identification of mantle plumes in the Emeishan Large Igneous Province. Episodes, 30: 32-42.

Xu, Y.-G., Luo, Z.-Y., Huang, X.-L., He, B., Xiao, L., Xie, L.-W., Shi, Y.-R., 2008. Zircon U–Pb and Hf isotope constraints on crustal melting associated with the Emeishan mantle plume. Geochimica et Cosmochimica Acta, 72: 3084-3104.

Xu, Y.-G., Wei, X., Luo, Z.-Y., Liu, H.-Q., Cao, J., 2014. The Early Permian Tarim Large Igneous Province: Main characteristics and a plume incubation model. Lithos, 204(3): 20-35.

Yang, S.-F., Chen, H.-L., Dong, C.-W., Jia, C.-Z., Wang, Z.-G., 1996. The discovery of Permian syenite inside Tarim basin and its geodynamic significance. Geochemistry, 25: 121-128 (in Chinese with English abstract).

Yang, S.-F., Chen, H.-L., Ji, D.-W., Li, Z.-L., et al., 2005. Geological process of Early to Middle Permian magmatism in Tarim Basin and its geodynamic significance. Geological Journal of China Universities, 11(4): 504-511 (in Chinese with English abstract).

Yang, S.-F., Li, Z.-L., Chen, H.-L., Chen, W., Yu, X., 2006a. $^{40}Ar-^{39}Ar$ dating of basalts from Tarim Basin, NW China and its implication to a Permian thermal tectonic event. Journal of Zhejiang University-Science A, 7(Suppl. II): 320-324.

Yang, S.-F., Li, Z.-L., Chen, H.-L., Dong, C.-W., Yu, X., Jia, C.-Z., Wei, G.-Q., 2006b. Permian large volume basalts in Tarim Basin. June 2006 LIP of the Month. http://www.largeigneousprovinces.org/06jun.

Yang, S.-F., Li, Z.-L., Chen, H.-L., Santosh, M., Dong, C.-W., Yu, X., 2007a. Permian bimodal dike of Tarim Basin, NW China: Geochemical characteristics and tectonic implications. Gondwana Research, 12(1-2): 113-120.

Yang, S.-F., Yu, X., Chen, H.-L., Li, Z.-L., Wang, Q.-H., Luo, J.-C., 2007b. Geochemical characteristics and petrogenesis of Permian Xiaohaizi ultrabasic dike in Bachu area, Tarim Basin. Acta Petrologica Sinica, 23: 1087-1096 (in Chinese with English abstract).

Yang, S.-F., Chen, H.-L., Li, Z.-L., Li, Y.-Q., Yu, X., Li, D.-X., Meng, L.-F., 2013. Early Permian Tarim Large Igneous Province in northwest China. Science China, Series D: Earth Sciences, 56(12): 2015-2026.

Yu, X., Chen, H.-L., Yang, S.-F., Li, Z.-L., Wang, Q.-H., Lin, X.-B., Xu, Y., Luo, J.-C., 2009. Geochemical features of Permian basalts in Tarim Basin and compared with Emeishan LIP. Acta Petrologica Sinica, 25(6): 1492-1498 (in Chinese with English abstract).

Yu, X., Yang, S.-F., Chen, H.-L., Chen, Z.-Q., Li, Z.-L., Batt, G.E., Li, Y.-Q., 2011. Permian flood basalts from the Tarim Basin, Northwest China: SHRIMP zircon U–Pb dating and geochemical characteristics. Gondwana Research, 20(2-3): 485-497.

Zhang, C.-L., Li, X.-H., Li, Z.-X., Ye, H.-M., Li, C.-N., 2008. A Permian layered intrusive complex in the western Tarim Block, northwestern China: Product of a ca. 275-Ma mantle plume. Journal of Geology, 116(3): 269-287.

Zhang, C.-L., Xu, Y.-G., Li, Z.-X., Wang, H.-Y., Ye, H.-M., 2010. Diverse Permian magmatism in the Tarim Block, NW China: Genetically linked to the Permian Tarim mantle plume. Lithos, 119(3-4): 537-552.

Zhang, D.-Y., Zhou, T.-F., Yuan, F., Jowitt, S.M., Fan, Y., Liu, S., 2012. Source, evolution and emplacement of Permian Tarim Basalts: Evidence from U–Pb dating, Sr–Nd–Pb–Hf isotope systematics and whole rock geochemistry of basalts from the Keping area, Xinjiang Uygur Autonomous Region, Northwest China. Journal of Asian Earth Sciences, 49: 175-190.

Zhang, D.-Y., Zhang, Z.-C., Santosh, M., Cheng, Z.-G., Huang, H., Kang, J.-L., 2013. Perovskite and baddeleyite from kimberlitic intrusions in the Tarim Large Igneous Province signal the onset of an end-Carboniferous mantle plume. Earth and Planetary Science Letters, 361: 238-248.

Zhang, H.-A., Li, Y.-J., Wu, G.-Y., Su, W., Qian, Y.-X., Meng, Q.-L., Cai, X.-Y., Han, L.-J., Zhao, Y., Liu, Y.-L., 2009. Isotopic geochronology of Permian igneous rocks in the Tarim Basin. Chinese Journal of Geology, 44(1): 137-158 (in Chinese with English abstract).

Zhang, S.-B., Ni, Y.-N., Gong, F.-H. et al., 2003. A Guide to the Stratigraphic Investigation on the Periphery of the Tarim Basin. Beijing: Petroleum Industry Press, p. 280.

Zhang, Y.-T., Liu, J.-Q., Guo, Z.-F., 2010. Permian basaltic rocks in the Tarim Basin, NW China: Implications for plume–lithosphere interaction. Gondwana Research, 18(4): 596-610.

Zhong, H., Zhou, X.-H., Zhou, M.-F., Sun, M., Liu, B.-G., 2002. Platinum-group element geochemistry of the Hongge layered intrusion in the Panxi area, southwestern China. Mineralium Deposita, 37: 226-239.

Zhou, D.-W., Liu, Y.-Q., Xin, X.-J., Hao, J.-R., Dong, Y.-P., Ouyang, Z.-J., 2006. Formation of the Permian basalts and implications of geochemical tracing for paleo-tectonic setting and regional tectonic background in the Turpan–Hami and Santanghu basins, Xinjiang. Science China, Series D, 49: 584-596.

Zhou, M.-F., Malpas, J.G., Song, X.-Y., et al., 2002. A temporal link between the Emeishan Large Igneous Province (SW China) and the end-Guadalupian mass extinction. Earth and Planetary Science Letters, 196(3-4): 113-122.

Zhou, M.-F., Robinson, P.T., Lesher, C.M., Keays, R.R., Zhang, C.-J., Malpas, J.G., 2005. Geochemistry, petrogenesis, and metallogenesis of the Panzhihua gabbroic layered intrusion and associated Fe–Ti–V oxide deposits, Sichuan Province, SW China. Journal of Petrology, 46: 2253-2280.

Zhou, M.-F., Arndt, N.T., Malpas, J., Wang, C.-Y., Kennedy, A.K., 2008. Two magma series and associated ore deposits types in the Permian Emeishan Large Igneous Province, SW China. Lithos, 103: 352-368.

Zhou, M.-F., Zhao, J.-H., Chang, Y.-J., Gao, J.-F., Wang, W., Yang, S.-H., 2009. OIB-like, heterogeneous mantle sources of Permian basaltic magmatism in the western Tarim Basin,

NW China: Implications for a possible Permian Large Igneous Province. Lithos, 113: 583-594.

Zhu, W.-B., Zheng, B.-H., Shu, L.-S., Ma, D.-S., Wu, H.-L., Li, Y.-X., Huang, W.-T., Yu, J.-J., 2011. Neoproterozoic tectonic evolution of the Precambrian Aksu blueschist terrane, northwestern Tarim, China: Insights from LA-ICP-MS zircon U–Pb ages and geochemical data. Precambrian Research, 185(3-4): 215-230.

Zou, S.-Y., Li, Z.-L., Song, B., Ernst, R.E., Li, Y.-Q., Ren, Z.-Y., Yang, S.-F., Chen, H.-L., Xu, Y.-G., Song, X.-Y., 2015. Zircon U–Pb dating, geochemistry and Sr–Nd–Pb–Hf isotopes of the Wajilitag alkali mafic dikes, and associated diorite and syenitic rocks: Implications for magmatic evolution of the Tarim Large Igneous Province. Lithos, 212-215: 428-442.

3

Geochemical Features of the Tarim LIP Rocks and Implications for the Magma Evolution

Abstract: The various igneous rocks in the Tarim LIP can be divided into 5 major units according to their geochemical and isotopic characteristics, including 3 earlier erupted basalt groups and 2 later emplaced intrusive rock units. There is an evolutional trend of the Sr–Nd–Hf isotopes from enriched mantle components in the earlier basalts towards depleted mantle components in the later intrusive rocks. The 3 basalt groups in the Tarim LIP have experienced variable degrees of crustal contamination, whereas the intrusive rocks are generally not contaminated. After removing the interference of crustal contamination, the uncontaminated parental magmas of various Tarim LIP rocks exhibit a wide range of $\varepsilon_{Nd}(t)$ values (ca. −5 to 5), reflecting source isotopic heterogeneity, which may be a consequence of plume–lithosphere interaction during the generation of the Tarim LIP.

Keywords: Geochemical features; Sr–Nd–Hf isotopes; Crustal contamination; Magma evolution; Three basalt groups

Based on their geochemical and isotopic characteristics, the various igneous rocks in the Tarim LIP can be divided into 5 units, which include 3 earlier erupted basalt groups and 2 later emplaced intrusive rock units (Fig. 3.1). This chapter will focus on the geochemical features of these main rock units and discuss the magma evolution of the Tarim LIP from a geochemical aspect.

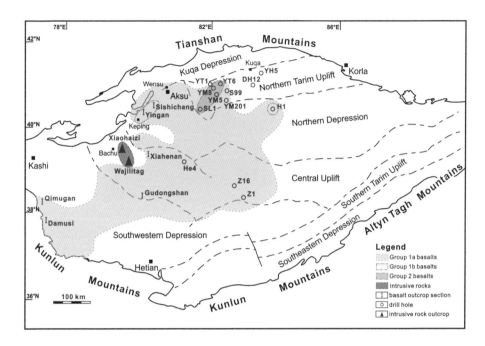

Fig. 3.1 The distributions of three basalt groups and Bachu intrusive rocks in the Tarim LIP (modified after Li et al., 2014). The locations of studied outcrops and boreholes are from Li Z.-L. et al. (2008; 2012), Li H.-Y. et al. (2013), Tian et al. (2010), Yu et al. (2011) and Zhang Y.-T. et al. (2010)

3.1 General Geochemical Features of the Tarim LIP Rocks

Generally, the various Tarim LIP rocks together cover a large range of SiO_2 (Fig. 3.2) from less than 45 wt.% (ultramafic rocks) to larger than 65 wt.% (felsic rocks). Like many other LIPs around the world, the intermediate rocks (e.g., andesite or diorite) are rare in the LIP. From ultramafic to mafic and then to felsic rocks, MgO, $Fe_2O_3^T$ and TiO_2 generally increase with a decrease in SiO_2, whereas Na_2O and K_2O decrease with a decrease in SiO_2. As shown in Fig. 3.2, most Tarim LIP rock samples exhibit good linear correlations on the major elemental binary diagram.

Apart from the major elements, the Tarim LIP rocks also exhibit similar patterns on the spidergram showing primitive mantle normalized trace elements (Fig. 3.3), which are generally also similar to many LIP rocks around the world. Nevertheless, the slight Nb depletion for the basalts indicates that the Tarim

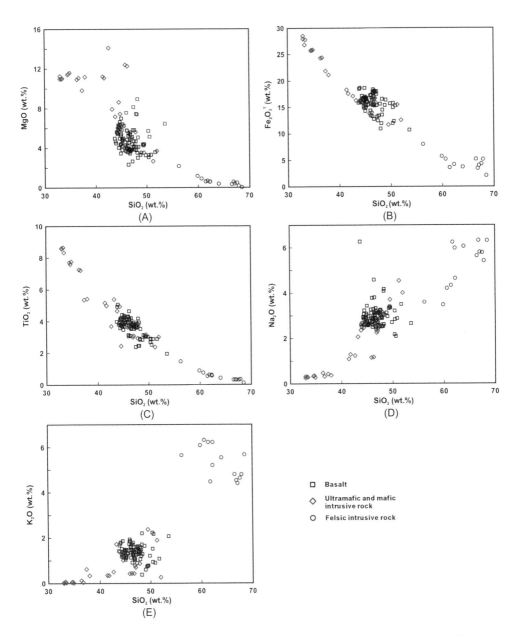

Fig. 3.2 Binary Diagrams of SiO_2 vs. MgO, $Fe_2O_3^T$, TiO_2, Na_2O and K_2O for the main Tarim LIP rocks. Original data and references are listed in Supplementary Table I. To minimize the effect of alteration, only data with LOI (loss on ignition) less than 2 wt.% are plotted here

basalts probably suffered crustal contamination during their eruption, and the depletion of the heavy rare earth elements (HREE) in the ultramafic and mafic intrusive rocks compared to the basalts suggests that the former probably derived from a deeper mantle source than the latter.

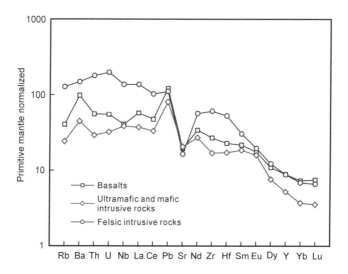

Fig. 3.3 Spidergram showing primitive mantle normalized trace elements for the average values of the main Tarim LIP rocks. Original data and references are listed in Supplementary Table I. To minimize the effect of alteration, only data with LOI (loss on ignition) less than 2 wt.% are plotted here

3.2 Geochemical Features of the Three Basalt Groups

The Tarim continental flood basalts (CFBs), as the dominant part of the Tarim LIP (Fig. 3.1), are mostly high-Ti basalts (TiO_2 >2.5 wt.% and Ti/Y >500) belonging to the alkaline series (Table 3.1). In general, they exhibit significant enrichments of large-ion lithophile and light rare-earth elements on the primitive mantle-normalized spidergram, which are similar to the ocean island basalts (OIBs) except for the slight Nb–Ta depletion (Fig. 3.4). Their high Ti/Y and Zr/Y ratios support a within-plate petrogenetic affinity.

Table 3.1 Geochemical and Sr–Nd–Hf isotopic characteristics of three basalt groups in the Tarim LIP

	Group 1a basalts	Group 1b basalts	Group 2 basalts
SiO_2 (wt.%)	44.14–48.14	46.70–51.62	47.39–54.55
TiO_2 (wt.%)	3.67–5.12	2.92–4.23	1.98–3.22
MgO (wt.%)	3.50–6.48	3.38–4.21	5.22–9.07
FeO_T (wt.%)	11.62–18.50	13.96–17.04	9.82–13.44
Na_2O+K_2O (wt.%)	3.74–7.75	3.83–5.14	3.03–6.56
Mg#	27–47	27–31	41–57
Ti/Y	344–708	470–628	448–1102
Nb/Yb	6.3–11.5	7.0–8.9	15.6–21.3
Th/Nb	0.10–0.20	0.20–0.29	0.12–0.18
Nb/U	23.6–42.4	14.4–20.4	24.3–60.3
Ce/Pb	3.3–17.8	7.8–11.5	15.6–23.7
$^{87}Sr/^{86}Sr_i$	0.7061–0.7080	0.7072–0.7085	0.7050–0.7053
$\varepsilon_{Nd}(t)$	−9.15 to −1.83	−4.14 to −1.79	−2.54 to 0.22
$\varepsilon_{Hf}(t)$	−2.81 to 2.11	−2.02 to −0.31	0.52–1.43

Note: Data are compiled from Supplementary Table I, only data with LOI (loss on ignition) less than 1.5 wt.% (<3.0 wt.% for Group 2 basalts due to the limited samples) are counted for major and trace elements as well as Sr isotopes here. $^{87}Sr/^{86}Sr_i$ represents the initial $^{87}Sr/^{86}Sr$ isotopic ratio. $\varepsilon_{Nd}(t)$ and $\varepsilon_{Hf}(t)$ are parts per 10^4 deviation of the initial $^{143}Nd/^{144}Nd$ and $^{176}Hf/^{177}Hf$ isotopic ratios between the sample and the chondritic uniform reservoir (CHUR), respectively

As shown in Fig. 3.4, the Tarim CFBs are primarily classified into two groups based on their geochemical distinctions, i.e., Groups 1 and 2 (Li Z.-L. et al., 2012). Group 1 basalts are mainly represented by those from the outcrop sections in the Keping area (e.g., Yingan, Kaipaizileike and Sishichang) and are widely distributed within the Tarim Basin according to the published and authors' unpublished data (Supplementary Table I), whereas Group 2 basalts by far are only identified from four boreholes (SL1, YM5, YM8 and YT6) located in the Northern Tarim Uplift (Fig. 3.1).

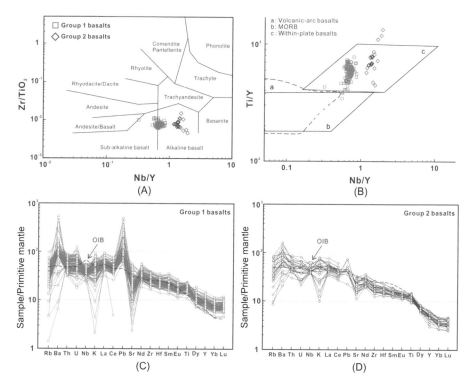

Fig. 3.4 (A) Nb/Y vs. ZrTiO$_2$ (after Winchester and Floyd, 1977) discrimination plots showing that most basalts (Groups 1 and 2) in the Tarim LIP belong to the alkaline series; (B) Nb/Y vs. Ti/Y (modified after Pearce, 1982) discrimination plots showing that the Tarim CFBs generally belong to the within-plate basalts; (C, D) Primitive mantle-normalized spidergram of the basalts in Groups 1 (C) and 2 (D), and a composite OIB as the reference (the dashed red line). The basalt data and their sources are given in Supplementary Table I; major elements are normalized to volatile-free compositions; both the primitive mantle normalizing values and the OIB data are from Sun and McDonough (1989)

3.2.1 Groups 1a and 1b Basalts

The Group 1 basalts in the Keping area, also known as Keping basalts in the literature (e.g., Li et al., 2008; Wei et al., 2014; Yu et al., 2012), can be further subdivided into Groups 1a and 1b, in the upper Kaipaizileike and lower Kupukuziman formations, respectively (Fig. 3.5; Li Z.-L. et al., 2012). These two sub-groups have many similar petrological and geochemical characteristics (Yu et al., 2011), but the earlier erupted Group 1b basalts show higher Th/Nb ratios (mostly >0.2; Fig. 3.6), and exhibit fairly low Nb/U (<21) and Ce/Pb (<12) ratios (Table 3.1), indicating more crustal contamination than the basalts in the latter Group 1a.

Further geochemical comparisons reveal that most samples from other studied outcrop sections (Xiahenan, Gudongshan, Damusi and Qimugan) and boreholes (S99, He4 and Z1) in the Tarim LIP could be incorporated into Group 1a basalts, yet only the samples from the H1 borehole in the Northern Depression resemble Group 1b basalts (Fig. 3.1; Li Z.-L. et al., 2012; Li Y.-Q. et al., 2014).

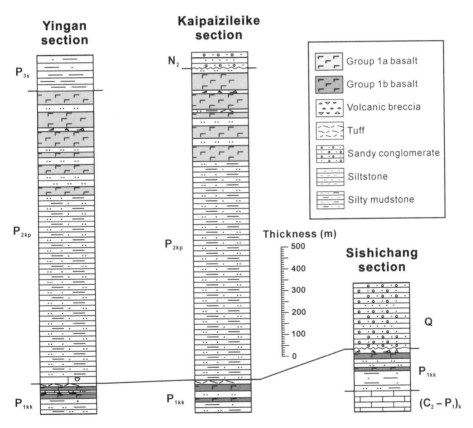

$(C_2-P_1)_k$: Kangkelin Formation; P_{1kk}: Kupukuziman Formation; P_{2kp}: Kaipaizileike Formation; P_{3s}: Shajingzi Formation; N_2: Kuche Formation; Q: Xiyu Formation

Fig. 3.5 Columnar sections of the Yingan, Kaipaizileike and Sishichang sections showing the distribution of different Keping basalt units in the Permian stratigraphy (modified after Yu et al., 2011)

3.2.2 The Group 2 Basalts

Compared to Group 1 basalts, the Group 2 basalts in the Northern Tarim Uplift show a clear depletion in the heavy rare-earth elements (beneath the OIB in the lower right spidergram in Fig. 3.4) and they have higher Nb/Yb ratios (≥15) that approach those of OIB (Fig. 3.6A). These geochemical features indicate that Group 2 basalts are more similar to OIB in composition than the basalts in both Groups 1a and 1b and the former also appears to be less contaminated by crustal materials (Fig. 3.6B).

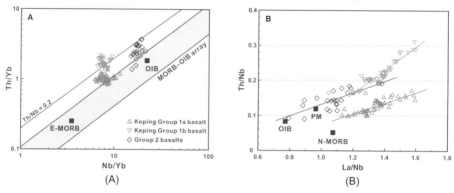

Fig. 3.6 Plots of (A) Nb/Yb vs. Th/Yb and (B) La/Nb vs. Th/Nb for Group 1a and 1b basalts in the Keping area and Group 2 basalts in the Northern Tarim Uplift. Original data can be found in Supplementary Table I. The compositions of PM (primitive mantle), N-MORB (normal mid-ocean ridge basalts), E-MORB (enriched mid-ocean ridge basalts) and OIB are from Sun and McDonough (1989), and the field of MORB–OIB array is taken from Pearce (2008)

Differences between Group 1 and Group 2 basalts can also be discerned from their distinctive Sr and Nd isotopic characteristics (Table 3.1), suggesting heterogeneous source compositions and variable crustal contamination during the generation of the Tarim CFBs. Despite the lack of a direct temporal sequence, Group 2 basalts are believed to be formed later than the Group 1 lavas, based on a progression of Sr–Nd–Hf isotopic compositions from the older basalts towards the younger intrusive rocks in the Tarim LIP (Fig. 3.7), but this hypothesis remains to be tested by future studies.

3.3 Geochemical Features of the Bachu Intrusive Rocks

The intrusive rocks in the Tarim LIP, mostly exposed at Xiaohaizi and Wajilitag of the Bachu area (Fig. 3.1), include ultramafic–mafic layered intrusions, syenitic rock bodies (and perhaps some A-type granitic plutons), radiating mafic dike swarms, and ultramafic and bimodal dikes (Li Y.-Q. et al., 2010, 2012; Li Z.-L. et al., 2012; Yang et al., 2007a, b; Zhang C.-L. et al., 2008, 2010; Zhou et al., 2009). Their ages range from 284 to 274 Ma, and thus are regarded as being substantially younger than the Tarim CFBs (see Chapter 2 for more details about their temporal relationships). Table 3.2 shows the geochemical and Sr–Nd–Hf isotopic characteristics of the different intrusive rocks in the Bachu area.

Table 3.2 Geochemical and Sr–Nd–Hf isotopic characteristics of the different Bachu intrusive rocks

	Ultramafic rocks	Mafic rocks	Syenitic rocks
SiO_2 (wt.%)	32.73–46.17	43.44–52.96	56.67–68.71
TiO_2 (wt.%)	1.60–8.65	2.38–5.41	0.11–1.47
MgO (wt.%)	7.44–22.55	2.69–7.96	0.01–2.19
FeO_T (wt.%)	12.34–25.41	10.30–15.44	1.85–7.25
Na_2O+K_2O (wt.%)	0.26–4.04	2.99–10.73	9.30–12.00
Mg#	43–72	26–49	1–35
Ti/Y	681–4337	227–1102	16–287
Nb/Yb	9.0–32.2	13.4–34.4	23.0–31.7
Th/Nb	0.01–0.12	0.06–0.22	0.10–0.21
Nb/U	27.7–54.5	21.2–67.4	16.9–35.1
Ce/Pb	13.7–22.5	11.3–30.8	28.0
$^{87}Sr/^{86}Sr_i$	0.7037–0.7051	0.7036–0.7068	0.7030–0.7041
$\varepsilon_{Nd}(t)$	1.17–3.96	−2.53 to 5.24	1.52–4.12
$\varepsilon_{Hf}(t)$	−0.28 to 3.99	−0.92 to 4.72	5.32–7.07

Note: Data are compiled from Supplementary Table I, some abnormal data (e.g., $^{87}Sr/^{86}Sr_i$ <0.7000) are not counted here

Studies on the different types of intrusive rocks in the Bachu area indicate that they were formed by magma differentiation (specifically, crystal fractionation and/or accumulation), probably from a common reservoir, and most have undergone negligible crustal contamination (Li Y.-Q. et al., 2012; Zhou et al., 2009; Zhang C.-L. et al., 2010). Most Bachu intrusive rocks exhibit OIB-like Sr–Nd–Hf isotopic signatures, clearly different from the three basalt groups (Fig. 3.7). Combined with their high melting temperature (up to 1300°C;

Zhou et al., 2009) and zircon-saturation temperatures (>800°C; Liu et al., 2013), the Bachu intrusive rocks reveal an affinity with a plume-derived (or plume-related) mantle source.

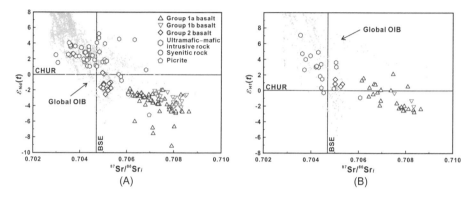

Fig. 3.7 Sr–Nd–Hf isotopic compositions of the three basalt groups, Bachu intrusive rocks and a picrite sample from the Northern Tarim Uplift. Data of the intrusive rocks in other areas (see Supplementary Table I) are not shown here for clarity. The original data and related calculations are given in Supplementary Table I. The $^{87}Sr/^{86}Sr$ ratio of Bulk Silicate Earth (BSE) is adopted from Allègre (2008). Small gray circles are the global OIB data from a compilation by Stracke (2012)

3.4 Effects of Crustal Contamination on the Tarim CFBs

Flood basalts that erupted on the continent are generally inevitably contaminated by crustal materials (Arndt and Christensen, 1992). The large variations of Th/Nb ratios and Sr–Nd–Hf isotopes among basalts from different groups (Figs. 3.6 and 3.7) suggest that the Tarim CFBs in different locations have undergone variable degrees of crustal contamination during eruptions. In this section, the potential sources of the crustal contamination on the basalts of the three groups are investigated and the effect of that contamination on them is evaluated.

3.4.1 Potential Sources of the Crustal Contamination During the Tarim CFB Eruptions

Zircon ($ZrSiO_4$) is a common accessory mineral in many igneous rocks, but it usually does not crystallize in the basaltic magmas (Poldervaart, 1956; Belousova et al., 2002; Hoskin and Schaltegger, 2003). It is notable that Keping basalts (both Groups 1a and 1b) contain zircons, some of which exhibit features typical of a magmatic origin and yielded SHRIMP and LA-ICP-MS U–Pb ages of ca. 290 Ma (Yu et al., 2011; Zhang D.-Y. et al., 2012), coincident with the ages of the formation of their host basalts. Further Hf isotope study on these Early Permian zircons shows that they generally possess lower $\varepsilon_{Hf}(t)$ values (−6.8 to −1.4) than the whole-rock $\varepsilon_{Hf}(t)$ values of their host basalts (−2.8 to 2.1), but are more similar to the zircons from a volcanic and pyroclastic rock (VPR) suite in the Lower Permian Xiaotikanlike Formation of the South Tianshan Orogen (Fig. 3.8), and thus are interpreted as xenocrysts that were originally derived from other contemporary igneous rocks in the South Tianshan Orogen (Li et al., 2014).

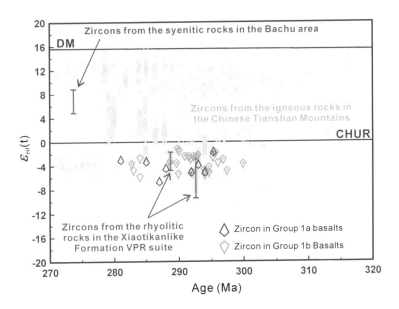

Fig. 3.8 Age vs. $\varepsilon_{Hf}(t)$ value plot of zircons from various Late Carboniferous to Early Permian (320–270 Ma) igneous rocks in the Tarim LIP and its adjacent Chinese Tianshan Mountains (Li et al., 2014). Groups 1a and 1b basalt symbols with crosses inside are marked for those data from Zhang D.-Y. et al. (2012)

Except those Early Permian zircons, the presence of other older inherited zircons (mostly >541 Ma) has also been reported in both of the basalt Groups 1a and 1b in the Keping area (Zhang D.-Y. et al., 2012). These Early Permian and Precambrian captured zircons provide direct evidence for crustal contamination during the eruptions of the Keping basalts, and hint that the components from the

Early Permian (ca. 290 Ma) igneous rocks in the South Tianshan Orogen (e.g., the Xiaotikanlike Formation VPR suite) and the Precambrian rocks from the basement of the Tarim Block are the sources of the potential contaminants.

Based on the information revealed from the zircons in the Keping basalts, three potential contaminant components are compiled (Table 3.3), which are: (1) the Tarim Precambrian rocks; (2) the Xiaotikanlike Formation volcanic and pyroclastic (VPR) suite rocks in the South Tianshan Orogen; (3) the high Th–U–Pb volcanic rocks in the Xiaotikanlike Formation VPR suite.

Table 3.3 Compiled compositions of components of the potential crustal contaminants of the Tarim CFBs

	(1)	(2)	(3)
Major elements (wt.%)			
SiO_2	62.41	70.04	75.49
TiO_2	0.75	0.58	0.25
Al_2O_3	15.65	13.95	12.91
$Fe_2O_3^T$	5.71	4.44	3.06
MgO	4.37	1.10	0.10
CaO	3.72	2.78	0.31
Na_2O	3.32	2.25	0.67
K_2O	2.44	4.51	6.90
Trace elements (ppm)			
Nb	7.14	30.5	157.0
La	32.8	64.0	246.0
Ce	64.7	119.0	343.0
Nd	28.8	52.3	136.0
Pb	12.1	28.2	106.0
Th	5.98	19.7	77.4
U	0.69	4.44	20.2
Trace element ratios			
Th/Nb	0.84	0.65	0.49
Nb/U	10.30	6.88	7.74
Ce/Pb	5.33	4.22	3.22
Nd isotopic composition			
$^{143}Nd/^{144}Nd_i$ ($T = 290$ Ma)	0.51135	0.51197	0.51200
$\varepsilon_{Nd}(t)$ ($T = 290$ Ma)	−17.9	−5.73	−5.25

Note: All the major and trace elements as well as $^{143}Nd/^{144}Nd_i$ ratios are average values; the trace element ratios and $\varepsilon_{Nd}(t)$ values are calculated based on the average values

The compiled values of the Tarim Precambrian rocks are averages of 103 Precambrian rock samples from the Tarim Block (data from Long et al., 2010, 2011; Zhang C.-L. et al., 2012a, b), including 34 mafic rock samples (e.g., gabbro and pyroxenite) and 69 felsic rock samples (mostly granitic rocks). This average composition is generally similar to the composition of the continental crust (Rudnick and Gao, 2003).

The compiled values of the Xiaotikanlike Formation VPR suite rocks are averages of 86 volcanic and pyroclastic rock samples in the Lower Permian Xiaotikanlike Formation from the northwestern part of Wensu County to the north of Kuche County (data from Liu D.-D. et al., 2014; Liu H.-Q. et al., 2014a, b; Li et al., 2009; Luo et al., 2008, 2013; Song et al., 2012; Wang et al., 2014), including 14 mafic rock samples (e.g., basalt and basaltic porphyrite) and 72 felsic rock samples (mainly rhyolite and dacite). The Xiaotikanlike Formation VPR suite are considered as the potential contaminants for the Tarim CFBs in the northern Tarim LIP (Li et al., 2014).

The compiled values of high Th–U–Pb volcanic rocks in the Xiaotikanlike Formation VPR suite are averages of 5 Xiaotikanlike Formation VPR suite rock samples with high Th, U and Pb concentrations (Th>50 ppm, U>10 ppm and Pb>80 ppm), including 2 rhyolitic ignimbrite samples (data from Liu H.-Q. et al., 2014b) and 3 quartz-albite porphyry samples (data from Liu D.-D. et al., 2014). They have much higher Th, U and Pb concentrations than other Xiaotikanlike Formation VPR suite rocks and the Tarim Precambrian rocks.

3.4.2 Evaluating the Effect of Crustal Contamination on the Tarim CFBs

To evaluate the crustal contamination effect on the Tarim CFBs in different locations, three trace element ratios (Th/Nb, Nb/U and Ce/Pb) and Nd isotopic compositions ($\varepsilon_{Nd}(t)$ value) are deployed in this section. As shown in Fig. 3.9, three basalt groups in different locations have distinguishable trends of magma evolution, indicating different processes during crustal contamination. Moreover, along each evolution trend tracking the basalt data back to the least contaminated region (e.g., MORB and OIB), all the data points will fall along a mixing line between a typical Bachu mafic dike sample (BC-5; Wei et al., 2014) and a picrite sample (YH5-1) from the Northern Tarim Uplift (Tian et al., 2010), which may represent the heterogeneous mantle source compositions during the generation of the Tarim LIPs (we will further discuss it in Section 3.5).

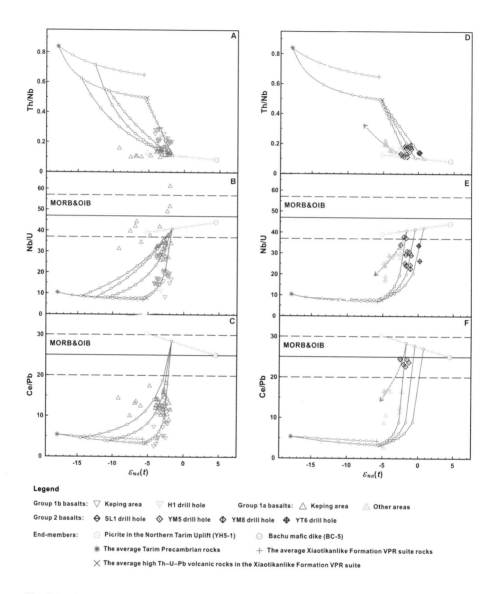

Fig. 3.9 Plots of $\varepsilon_{Nd}(t)$ value vs. Th/Nb (A and D), Nb/U (B and E) and Ce/Pb (C and F) ratios for the three basalt groups in different locations. Data of the Tarim basalts and two source region end-members (YH5-1 and BC-5) are from Supplementary Table I, whereas the data for the components of the potential crustal contaminants are given in Table 3.3. Small circles on the crustal contamination curves (red color) and end-member mixing lines (other colors) show the 10% intervals (see the text for details). The field of MORB and OIB, bounded by dashed lines, was adopted from Hofmann et al. (1986)

A further quantitative simulation suggests that the geochemical features of Group 1b basalts can generally be reproduced by a two-stage melting and contamination model. First, magmas are generated from a source characterized by two end-members, i.e., the picrite sample (YH5-1) from the Northern Tarim Uplift mixed with 30% of the Bachu mafic dike sample (BC-5), though the mixing proportion could vary with the selection of end-members and thus is not important here. These melts are then variably contaminated (ca. 5%–25%) predominantly by the high Th–U–Pb volcanic components from the Xiaotikanlike Formation VPR suite with 10% of the Tarim Precambrian rocks (Fig. 3.9A, B and C). Contamination of the melts parental to Group 1b basalts mainly by the high Th–U–Pb rock components would greatly increase the Th/Nb ratios, and decrease the Nb/U and Ce/Pb ratios without significantly changing the compositions of the major elements (Li et al., 2014).

Most of the Group 1a basalt samples in the Keping area could be simulated from the same source as that of Group 1b with less than 20% contamination by the same contaminant combinations. In this case, however, the proportion of high Th–U–Pb volcanic components would vary from 90% to less than 10% (Figs. 3.9A, B and C), indicating that they were not so important during the contamination process as the Group 1a magmas. The high Th–U–Pb volcanic components might also be replaced by the components of average Xiaotikanlike Formation VPR suite rocks with similar contamination effects when the Tarim Precambrian rocks are dominant (e.g., 70%) in the contaminants (the relevant simulated results are only shown in Figs. 3.9A and B for clarity). Note that some samples reported by Zhou et al. (2009) have compositional features that are similar to the other Group 1a basalts except for their much lower $\varepsilon_{Nd}(t)$ values (< –6.0; see Supplementary Table I), which require further evaluation and will not be discussed here. Most of the Group 1a basalt samples in other areas (e.g., central and southwestern Tarim Basin) also exhibit a clear trend towards the Tarim Precambrian rocks (Figs. 3.9D, E and F), implying contamination primarily by the latter. Detailed modeling on these basalts should be the focus of subsequent work.

Using the same modeling conditions, the Group 2 basalt samples in the Northern Tarim Uplift could also be produced by the same contaminants as those for Group 1b basalts, but from variable source compositions towards the Bachu mafic dike sample (Figs. 3.9D, E and F). In this scenario, the degree of crustal contamination is less than 10% for different Group 2 basalt samples and is less than 5% for those reported by Tian et al. (2010), consistent with their conclusion that Group 2 basalts have undergone little crustal contamination.

The modeling results are generally consistent with different discrimination plots (Fig. 3.9), although the "true" degree of contamination for the basalts in different localities would depend on the real compositions of the magma source and the contaminants. Nevertheless, the present modeling has shown that crustal contamination of the Tarim CFBs cannot simply be ruled out, especially for Group 1b basalts, and their effects deserve further investigation. The striking positive Th–U–Pb anomalies in Group 1b basalts (Fig. 3.10), about 1.5 times more than those of Group 1a, are the explicit marks of contamination by some high Th–U–Pb crustal materials. Around the Tarim LIP, such rocks are almost exclusively reported in the South Tianshan Orogen (e.g., Huang et al., 2014; Liu D.-D. et al., 2014; Liu H.-Q. et al., 2014b), which may also explain why Group 1b basalts are concentrated in the northern part of the Tarim LIP.

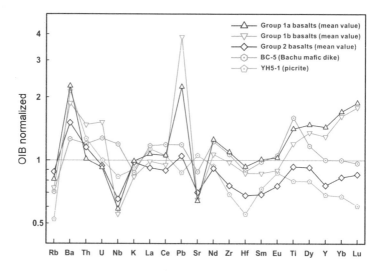

Fig. 3.10 Patterns of the abundance of OIB normalized elements for the three basalt groups (mean value), the Bachu mafic dike sample of BC-5 and the picrite sample of YH5-1. The original data are given in Supplementary Table I. The OIB normalizing values are from Sun and McDonough (1989)

3.5 Implications for Source Isotopic Heterogeneity and Plume–Lithosphere Interaction in the Tarim LIP

The picrite sample of YH5-1 (reported by Tian et al., 2010) used for modeling in subsection 3.3.2 yielded a negative $\varepsilon_{Nd}(t)$ value (−5.18) that is lower than most other samples in the Tarim LIP, but its MORB- and OIB-like Nb/U (39.0) and Ce/Pb (30.3) ratios exclude significant crustal contamination (Fig. 3.9). Therefore, the Nd isotopic signature of this picrite sample should reflect its source, which would be similar to the EM-I type enriched mantle reservoir (Hart et al., 1992) situated in the sub-continental lithospheric mantle (SCLM; Hofmann, 1997). The selected Bachu mafic dike sample of BC-5 (data from Wei et al., 2014), on the other hand, possesses typical OIB-type geochemical and Nd isotopic characteristics and is thought to be derived from the convecting asthenospheric mantle (Wei et al., 2014; Xu et al., 2014).

Despite the varying effects of crustal contamination on different group basalts, most of them retain $\varepsilon_{Nd}(t)$ values within the range identified for the end-member mantle sources (ca. −5 to 5; Supplementary Table I). The regression trends of different group basalts in Fig. 3.9 also suggest that their uncontaminated parental magmas would mostly fall along the mixing line between the above two samples. If not contaminated, Group 1 basalts would have relatively uniform $\varepsilon_{Nd}(t)$ values of ca. −1.7, whereas Group 2 basalts would display a wider range roughly between −1.7 and 0.8 (Fig. 3.9). Compared to the basalts, the less contaminated Bachu intrusive rocks generally exhibit positive $\varepsilon_{Nd}(t)$ values that are larger than 0.8. This, in addition with their similar highly variable but still well-correlated Hf and Sr isotopic compositions (Fig. 3.7), denotes the involvement of isotopically heterogeneous sources during the generation of the Tarim LIP.

Source isotopic heterogeneity is ubiquitous in the Earth's mantle (Stracke, 2012), and in the Tarim LIP this phenomenon could be ascribed to the plume–lithosphere interaction as will be illustrated in details in Chapter 4. The isotopically enriched picrites and most basalts (Group 1) were produced by partial melting of some enriched mantle components in the SCLM beneath the Tarim Block. This melting was due to conductive heating resulting from an incubating mantle plume (Xu et al., 2014), and was deeper for the picrites than the basalts as evidenced by the higher middle to heavy rare-earth element ratios in the former (Fig. 3.10). The younger and isotopically-depleted Bachu intrusive rocks were generated by a subsequent decompression melting of the mantle plume itself. Beyond this, it is worth noting that Group 2 basalts are only found in a small region in the Northern Tarim Uplift, but exhibit higher and more variable $\varepsilon_{Nd}(t)$ values than the widespread Group 1 basalts (both refer to the absence of crustal contamination). This may suggest that during the generation of Group 2 basalts, the upwelling mantle plume not only provided an enormous amount of heat, but it also continuously injected isotopically-depleted plume components into the isotopically-enriched magma source region in the SCLM.

References

Allègre, C.J., 2008. Isotope Geology. New York: Cambridge University Press.

Arndt, N.T., Christensen, U., 1992. The role of lithospheric mantle in continental flood volcanism: Thermal and geochemical constraints. Journal of Geophysical Research: Solid Earth, 97(B7): 10967-10981.

Belousova, E.A., Griffin, W.L., O'Reilly, S.Y., Fisher, N.I., 2002. Igneous zircon: Trace element composition as an indicator of source rock type. Contributions to Mineralogy and Petrology, 143(5): 602-622.

Hart, S.R., Hauri, E.H., Oschmann, L.A., Whitehead, J.A., 1992. Mantle plumes and entrainment: Isotopic evidence. Science, 256(5056): 517-520.

Hofmann, A.W., Jochum, K.P., Seufert, M., White, W.M., 1986. Nb and Pb in oceanic basalts: New constraints on mantle evolution. Earth and Planetary Science Letters, 79(1-2): 33-45.

Hofmann, A.W., 1997. Mantle geochemistry: The message from oceanic volcanism. Nature, 385(6613): 219-229.

Hoskin, P.W.O., Schaltegger, U., 2003. The composition of zircon and igneous and metamorphic petrogenesis. Reviews in Mineralogy and Geochemistry, 53(1): 27-62.

Huang, H., Zhang, Z.-C., Santosh, M., Zhang, D.-Y., 2014. Geochronology, geochemistry and metallogenic implications of the Boziguo'er rare metal-bearing peralkaline granitic intrusion in South Tianshan, NW China. Ore Geology Reviews, 61: 157-174.

Li, H.-Y., Huang, X.-L., Li, W.-X., Cao, J., He, P.-L., Xu, Y.-G., 2013. Age and geochemistry of the Early Permian basalts from Qimugan in the southwestern Tarim basin. Acta Petrologica Sinica, 29(10): 3353-3368 (in Chinese with English abstract).

Li, J.-B., Wei, Y.-F., Du, H.-X., Sun, T., 2009. Geochemical characteristics of the Lower Permian Xiaotikanlike Formation volcanic rocks in southwestern Tianshan. Xinjiang Geology, 27(4): 315-319 (in Chinese with English abstract).

Li, Y.-Q., Li, Z.-L., Sun, Y.-L., Chen, H.-L., Yang, S.-F., Yu, X., 2010. PGE and geochemistry of Wajilitag ultramafic cryptoexplosive brecciated rocks from Tarim Basin: Implications for petrogenesis. Acta Petrologica Sinica, 26(11): 3307-3318 (in Chinese with English abstract).

Li, Y.-Q., Li, Z.-L., Chen, H.-L., Yang, S.-F., Yu, X., 2012. Mineral characteristics and metallogenesis of the Wajilitag layered mafic–ultramafic intrusion and associated Fe–Ti–V oxide deposit in the Tarim Large Igneous Province, northwest China. Journal of Asian Earth Sciences, 49: 161-174.

Li, Y.-Q., Li, Z.-L., Yu, X., Langmuir, C.H., Santosh, M., Yang, S.-F., Chen, H.-L., Tang, Z.-L., Song, B., Zou, S.-Y., 2014. Origin of the Early Permian zircons in Keping basalts and magma evolution of the Tarim Large Igneous Province (northwestern China). Lithos, 204: 47-58.

Li, Z.-L., Yang, S.-F., Chen, H.-L., Langmuir, C.H., Yu, X., Lin, X.-B., Li, Y.-Q., 2008. Chronology and geochemistry of Taxinan basalts from the Tarim Basin: Evidence for

Permian plume magmatism. Acta Petrologica Sinica, 24(5): 959-970 (in Chinese with English abstract).

Li, Z.-L., Li, Y.-Q., Chen, H.-L., Santosh, M., Yang, S.-F., Xu, Y.-G., Langmuir, C.H., Chen, Z.-X., Yu, X., Zou, S.-Y., 2012. Hf isotopic characteristics of the Tarim Permian large igneous province rocks of NW China: Implication for the magmatic source and evolution. Journal of Asian Earth Sciences, 49: 191-202.

Liu, D.-D., Guo, Z.-J., Jolivet, M., Cheng, F., Song, Y., Zhang, Z.-Y., 2014. Petrology and geochemistry of Early Permian volcanic rocks in South Tian Shan, NW China: Implications for the tectonic evolution and Phanerozoic continental growth. International Journal of Earth Sciences, 103(3): 737-756.

Liu, H.-Q., Xu, Y.-G., He, B., 2013. Implications from zircon-saturation temperatures and lithological assemblages for Early Permian thermal anomaly in northwest China. Lithos, 182-183: 125-133.

Liu, H.-Q., Xu, Y.-G., Tian, W., Zhong, Y.-T., Mundil, R., Li, X.-H., Yang, Y.-H., Luo, Z.-Y., Shang-Guan, S.-M., 2014a. Origin of two types of rhyolites in the Tarim Large Igneous Province: Consequences of incubation and melting of a mantle plume. Lithos, 204: 59-72.

Liu, H.-Q., Xu, Y.-G., Zhong, Y.-T., Luo, Z.-Y., Mundil, R., 2014b. Post-collisional peraluminous volcanism in the South Tianshan Orogen: Petrogenesis and tectonic implications (Personal Communication).

Long, X.-P., Yuan, C., Sun, M., Zhao G.-C., Xiao, W.-J., Wang, Y.-J., Yang, Y.H., Hu, A.-Q., 2010. Archean crustal evolution of the northern Tarim craton, NW China: Zircon U–Pb and Hf isotopic constraints. Precambrian Research, 180(3-4): 272-284.

Long, X.-P., Yuan, C., Sun, M., Kröner, A., Zhao, G.-C., Wilde, S., Hu, A.Q., 2011. Reworking of the Tarim Craton by underplating of mantle plume-derived magmas: Evidence from Neoproterozoic granitoids in the Kuluketage area, NW China. Precambrian Research, 187 (1-2): 1-14.

Luo, J.-H., Che, Z.-C., Cao, Y.-Z., Zhang, J.-Y., 2008. Geochemical and geochronological characteristics and its tectonic significance of Early Permian acid volcanic rocks of Xiaotikanlike Formation in the southern margin of South Tianshan orogen, NW China. Acta Petrologica Sinica, 24(10): 2281-2288 (in Chinese with English abstract).

Luo, J.-H., Che, Z.-C., Zhou, N.-C., Yang, W., Song, H.-X., Hou, X.-J., Han, K., 2013. Geochemistry and isotopic geochronology of dacites from the Lower Permian Xiaotikanlike Formation in the southern margin of South Tianshan, and its tectonic significances. Acta Geologica Sinica, 87(1): 29-37 (in Chinese with English abstract).

Pearce, J.A., 1982. Trace element characteristics of lavas from destructive plate boundaries. In: Thorpe, R.S. (Ed.), Orogenic Andesites and Related Rocks. Chichester: John Wiley and Sons, pp. 528-548.

Pearce, J.A., 2008. Geochemical fingerprinting of oceanic basalts with applications to ophiolite classification and the search for Archean oceanic crust. Lithos, 100(1-4): 14-48.

Poldervaart, A., 1956. Zircon in rocks. 2. Igneous rocks. American Journal of Science, 254(9):

521-554.

Rudnick, R.L., Gao, S., 2003. Composition of the continental crust. In: Holland, H.D., Turekian, K.K. (Eds.), Treatise on Geochemistry, Volume 3: The Crust. Amsterdam: Elsevier, pp. 1-64.

Stracke, A., 2012. Earth's heterogeneous mantle: A product of convection-driven interaction between crust and mantle. Chemical Geology, 330-331: 274-299.

Song, H.-X., Luo, J.-H., Tang, J.-Y., Zhang, C., 2012. Geochemistry, and its tectonic significances of basaltic magma of Xiaotikanlike Formation in southern margin of south Tianshan, Baichen County. Chinese Journal of Geology, 47(3): 886-898 (in Chinese with English abstract).

Sun, S.-S., McDonough, W.F., 1989. Chemical and isotopic systematics of oceanic basalts: implications for mantle composition and processes. In: Saunders, A.D., Norry, M.J. (Eds.), Magmatism in the Ocean Basins. Geological Society, London: Special Publications, 42: 313-345.

Tian, W., Campbell, I.H., Allen, C.M., Guan, P., Pan W.-Q., Chen, M.-M., Yu, H.-J., Zhu, W.-P., 2010. The Tarim picrite–basalt–rhyolite suite, a Permian flood basalt from northwest China with contrasting rhyolites produced by fractional crystallization and anatexis. Contributions to Mineralogy and Petrology, 160(3): 407-425.

Wang, M., Zhang, J.-J., Qi, G.-W., Liu, J., 2014. Geochemistry and geochronology of Early Permian acid volcanic rocks along Kuqa River and its tectonic implication in the southern margin of South Tianshan Orogen Xinjiang. Chinese Journal of Geology, 49(1): 242-258 (in Chinese with English abstract).

Wei, X., Xu, Y.-G., Feng, Y.-X., Zhao J.-X., 2014. Plume–lithosphere interaction in the generation of the Tarim Large Igneous Province, NW China: Geochronological and geochemical constraints. American Journal of Science, 314: 314-356.

Winchester, J.A, Floyd, P.A., 1977. Geochemical discrimination of different magma series and their differentiation products using immobile elements. Chemical Geology, 20: 325-343.

Xu, Y.-G., Wei, X., Luo, Z.-Y., Liu, H.-Q., Cao, J., 2014. The Early Permian Tarim Large Igneous Province: Main characteristics and a plume incubation model. Lithos, 204: 20-35.

Yang, S.-F., Li, Z.-L., Chen, H.-L., Santosh, M., Dong, C.-W., Yu, X., 2007a. Permian bimodal dike of Tarim Basin, NW China: Geochemical characteristics and tectonic implications. Gondwana Research, 12(1-2): 113-120.

Yang, S.-F., Yu, X., Chen, H.-L., Li, Z.-L., Wang, Q.-H., Luo, J.-C., 2007b. Geochemical characteristics and petrogenesis of Permian Xiaohaizi ultrabasic dike in Bachu area, Tarim Basin. Acta Petrologica Sinica, 23(5): 1087-1096 (in Chinese with English abstract).

Yu, J.-C., Mo, X.-X., Yu, X.-H., Dong, G.-C., Fu, Q., Xing, F.-C., 2012. Geochemical characteristics and petrogenesis of Permian basaltic rocks in Keping area, Western Tarim Basin: A record of plume-lithosphere interaction. Journal of Earth Science, 23(4): 442-454.

Yu, X., Yang, S.-F., Chen, H.-L., Chen, Z.-Q., Li, Z.-L., Batt, G.E., Li, Y.-Q., 2011. Permian flood basalts from the Tarim Basin, Northwest China: SHRIMP zircon U–Pb dating and geochemical characteristics. Gondwana Research, 20 (2-3): 485-497.

Zhang, C.-L., Li, X.-H., Li, Z.-X., Lu, S.-N., Ye, H.-M., Li, H.-M., 2007. Neoproterozoic ultramafic–mafic-carbonatite complex and granitoids in Quruqtagh of northeastern Tarim Block, western China: Geochronology, geochemistry and tectonic implications. Precambrian Research, 152(3-4): 149-169.

Zhang, C.-L., Li, X.-H., Li, Z.-X., Ye, H.-M., Li, C.-N., 2008. A Permian layered intrusive complex in the Western Tarim Block, Northwestern China: Product of a ca. 275-Ma mantle plume? Journal of Geology, 116(3): 269-287.

Zhang, C.-L., Xu, Y.-G., Li, Z.-X., Wang, H.-Y., Ye, H.-M., 2010. Diverse Permian magmatism in the Tarim Block, NW China: Genetically linked to the Permian Tarim mantle plume? Lithos, 119(3-4): 537-552.

Zhang, C.-L., Li, H.-K., Santosh, M., Li, Z.-X., Zou, H.-B., Wang, H.-Y., Ye, H.-M., 2012a. Precambrian evolution and cratonization of the Tarim Block, NW China: Petrology, geochemistry, Nd-isotopes and U–Pb zircon geochronology from Archaean gabbro-TTG–potassic granite suite and Paleoproterozoic metamorphic belt. Journal of Asian Earth Sciences, 47: 5-20.

Zhang, C.-L., Zou, H.-B., Wang, H.-Y., Li, H.-K., Ye, H.-M., 2012b. Multiple phases of the Neoproterozoic igneous activity in Quruqtagh of the northeastern Tarim Block, NW China: Interaction between plate subduction and mantle plume. Precambrian Research, 222-223: 488-502.

Zhang, Y.-T., Liu, J.-Q., Guo, Z.-F., 2010. Permian basaltic rocks in the Tarim Basin, NW China: Implications for plume–lithosphere interaction. Gondwana Research, 18(4): 596-610.

Zhang, D.-Y., Zhou, T.-F., Yuan, F., Jowitt, S.M., Fan, Y., Liu, S., 2012. Source, evolution and emplacement of Permian Tarim Basalts: Evidence from U–Pb dating, Sr–Nd–Pb–Hf isotope systematics and whole rock geochemistry of basalts from the Keping area, Xinjiang Uygur Autonomous Region, Northwest China. Journal of Asian Earth Sciences, 49: 175-190.

Zhou, M.-F., Zhao, J.-H., Jiang, C.-Y., Gao, J.-F., Wang, W., Yang, S.-H., 2009. OIB-like, heterogeneous mantle sources of Permian basaltic magmatism in the western Tarim Basin, NW China: Implications for a possible Permian large igneous province. Lithos, 113(3-4): 583-594.

Supplementary Table I Characteristic geochemical and Sr-Nd-Hf isotopic compositions of different igneous rock units in the TLIP

Rock unit	Sample No.	Locality	SiO_2 (wt.%)	TiO_2 (wt.%)	MgO (wt.%)	FeO_t (wt.%)	Na_2O+K_2O (wt.%)	Zr (ppm)	Mg#	Ti/Y	Zr/TiO_2	Nb/Y	Th/Yb	Nb/Yb	Th/Nb	La/Nb	Nb/U	Ce/Pb	Sm/Yb	La/Sm	$^{87}Sr/^{86}Sr_i$	$\varepsilon_{Nd}(t)$	T_{DM}^{Nd} (Ma)	$\varepsilon_{Hf}(t)$	T_{DM}^{Hf} (Ma)	Data source
Basalts	AQ1-2	KaFK	45.20	3.63	6.40	14.51	4.39	291	44	481	80	0.63	1.1	9.9	0.11	1.26	34.0	12.8	3.24	3.84	0.7071	−2.53	1443	-	-	Zhou et al. (2009)
	AQ1-5	KaFK	44.82	4.02	5.55	15.06	4.82	321	40	479	80	0.73	1.0	10.0	0.10	1.13	38.8	12.3	2.96	3.81	-	-	-	-	-	Zhou et al. (2009)
	AQ2-1	KaFK	44.98	3.95	5.50	14.94	4.52	285	40	488	72	0.65	1.0	9.1	0.11	1.23	31.7	8.5	2.99	3.75	-	-	-	-	-	Zhou et al. (2009)
	AQ2-6	KaFK	44.87	3.87	5.71	14.77	4.54	266	41	492	69	0.62	0.9	8.0	0.12	1.29	33.7	10.7	2.86	3.64	-	-	-	-	-	Zhou et al. (2009)
	AQ3-3	KaFK	45.73	3.85	5.91	13.64	3.99	296	44	465	77	0.59	0.9	8.4	0.10	1.29	34.5	12.4	3.48	3.13	0.7063	−6.02	1729	-	-	Zhou et al. (2009)
	AQ3-8	KaFK	45.42	3.79	6.48	13.81	4.12	284	46	525	75	0.70	0.8	8.5	0.10	1.17	39.7	10.0	2.52	3.91	0.7071	−7.55	1919	-	-	Zhou et al. (2009)
	AQ3-10	KaFK	45.92	3.53	7.13	13.08	3.70	307	49	485	87	0.62	0.9	7.9	0.12	1.33	38.0	10.6	2.64	3.99	-	-	-	-	-	Zhou et al. (2009)
	AQ4-2	KaFK	44.44	4.21	3.69	14.62	4.61	270	31	570	64	0.65	0.9	8.4	0.11	1.21	42.0	13.1	2.63	3.86	0.7071	−6.83	1774	-	-	Zhou et al. (2009)
	AQ5-2	KaFK	45.99	3.92	2.83	12.66	6.74	280	28	578	71	0.67	0.9	8.6	0.11	1.28	44.2	13.5	3.37	3.26	0.7080	−6.68	1732	-	-	Zhou et al. (2009)
	AQ6-2	KaFK	47.32	3.94	3.50	11.62	4.53	277	35	571	70	0.74	0.9	8.8	0.10	1.17	41.8	13.2	2.57	4.02	0.7065	−3.44	1487	-	-	Zhou et al. (2009)
	AQ7-2	KaFK	46.61	4.08	2.13	13.79	6.53	299	22	490	73	0.64	1.0	8.3	0.12	1.33	34.1	10.2	2.55	4.29	0.7074	−3.17	1454	-	-	Zhou et al. (2009)
	AQ7-4	KaFK	44.68	3.61	2.34	11.98	7.04	259	26	514	72	0.61	1.0	8.1	0.12	1.35	46.4	10.9	3.01	3.67	-	-	-	-	-	Zhou et al. (2009)
	AQ8-2	KaFK	48.01	3.76	5.87	12.16	4.29	296	46	516	79	0.71	1.3	8.5	0.16	1.53	31.3	14.3	2.58	5.06	0.7079	−9.15	2155	-	-	Zhou et al. (2009)
	AQ8-8	KaFK	45.63	3.75	6.76	13.69	3.80	276	47	551	74	0.75	1.0	9.6	0.11	1.32	37.4	17.4	2.74	4.61	0.7074	−4.74	1658	-	-	Zhou et al. (2009)
	AQY1-2	KaFK	45.40	4.23	6.19	14.59	4.07	357	43	564	84	0.86	1.0	10.3	0.10	1.28	45.0	16.2	2.94	4.51	-	-	-	-	-	Zhou et al. (2009)
	AQY2-2	KaFK	46.23	4.00	6.35	14.12	4.15	337	45	526	84	0.86	1.2	11.5	0.10	1.20	42.4	17.8	3.10	4.44	-	-	-	-	-	Zhou et al. (2009)
	AQY4-4	KaFK	47.72	3.85	5.86	11.99	4.02	280	47	522	73	0.64	1.1	7.7	0.15	1.59	28.8	15.5	2.54	4.80	-	-	-	-	-	Zhou et al. (2009)
	kp13-4b	KaFK	47.13	3.12	2.38	12.13	6.39	262	26	599	84	0.73	1.2	7.2	0.16	1.45	26.0	10.3	2.60	4.02	0.7064	−2.64	1380	−0.8	1162	1*
	yg12-1a	KaFY	47.19	3.89	5.71	13.01	4.21	261	44	648	67	0.65	1.0	6.4	0.15	1.39	26.1	11.6	2.42	3.67	0.7068	−2.90	1507	-	-	Yu (2009)
Group 1a basalts	Yg0512-3b	KaFY	45.43	3.67	6.42	14.51	4.15	237	44	680	65	0.67	0.9	6.4	0.14	1.34	27.1	10.3	2.29	3.75	0.7062	−2.26	1362	1.6	1073	2*
	Yg0512-4a	KaFY	44.29	3.54	3.80	12.79	6.42	224	35	717	63	0.66	0.9	6.6	0.14	1.43	51.6	9.0	2.61	3.60	0.7075	−2.03	1410	2.1	1028	2*
	Yg0512-4k	KaFY	44.85	3.94	5.59	14.89	4.14	272	40	705	69	0.72	0.9	7.4	0.11	1.29	33.8	11.5	2.50	3.82	0.7061	−1.83	1302	1.3	1056	2*
	Yg0512-5c	KaFY	46.56	3.90	5.28	14.80	4.08	294	39	597	75	0.67	1.0	6.9	0.15	1.50	29.1	10.7	2.58	3.97	0.7069	−3.35	1413	−1.9	1234	2*
	Yg0512-5f	KaFY	45.30	3.97	6.01	15.24	4.27	291	41	663	73	0.71	1.1	7.8	0.14	1.37	30.8	11.5	2.76	3.88	0.7067	−2.80	1401	0.8	1056	2*

(To be continued)

(Supplementary Table I)

Rock unit	Sample No.	Locality	SiO$_2$ (wt.%)	TiO$_2$ (wt.%)	MgO (wt.%)	FeO$_t$ (wt.%)	Na$_2$O+K$_2$O (wt.%)	Zr (ppm)	Mg#	Ti/Y	Zr/TiO$_2$	Nb/Y	Th/Yb	Nb/Yb	Th/Nb	La/Nb	Nb/U	Ce/Pb	Sm/Yb	La/Sm	^{87}Sr/^{86}Sr$_i$	ε$_{Nd}(t)$	T$_{DM}^{Nd}$ (Ma)	ε$_{Hf}(t)$	T$_{DM}^{Hf}$ (Ma)	Data source
Group 1a basalts	Yg0512-6b	KaFY	45.25	4.18	5.03	15.30	4.31	272	37	708	65	0.73	1.0	7.6	0.14	1.40	31.3	11.8	2.73	3.91	0.7068	-3.92	1448	-0.3	1145	2*
	Yg0512-6c	KaFY	44.99	4.17	5.95	15.64	4.31	307	40	641	74	0.73	1.1	7.8	0.14	1.40	31.1	12.5	2.86	3.80	0.7067	-2.99	1412	0.8	1071	2*
	Yg0512-7a	KaFY	42.26	5.19	5.34	18.00	4.60	381	35	633	73	0.75	1.0	8.1	0.13	1.38	49.1	18.1	2.87	3.86	0.7066	-2.39	1332	-0.5	1139	2*
	Yg0512-7f	KaFY	42.56	5.01	3.20	17.87	4.85	353	24	639	70	0.72	1.0	7.7	0.12	1.42	32.9	10.7	2.86	3.83	0.7070	-2.75	1374	0.1	1135	2*
	Yg0512-8d	KaFY	45.84	3.72	3.54	17.13	4.53	421	27	407	113	0.66	1.1	7.2	0.15	1.53	28.3	12.5	2.71	4.06	0.7077	-3.86	1397	-2.3	1238	2*
	Yg0512-8i	KaFY	44.80	3.77	6.22	15.09	4.41	258	42	675	68	0.70	0.9	7.3	0.13	1.38	31.9	11.7	2.56	3.90	0.7065	-2.41	1330	0.3	1119	2*
	Yg07	KaFY	43.00	4.20	4.52	16.24	6.43	288	33	779	69	0.80	1.0	8.6	0.12	1.14	61.2	13.1	2.72	3.63	0.7075	-2.01	1300	-	-	Li Y.-Q. et al. (2012)
	K-01	KaFS	51.65	3.25	4.45	18.99	4.24	262	29	407	81	0.64	1.1	7.1	0.16	1.59	26.2	9.4	2.82	4.01	-	-	-	-	-	Zhang et al. (2012)
	K-02	KaFS	46.00	4.37	3.20	14.89	5.20	442	28	434	101	0.32	1.2	3.7	0.32	3.31	16.2	15.2	2.99	4.04	-	-	-	-	-	Zhang et al. (2012)
	K-03	KaFS	46.23	4.29	3.76	18.50	4.84	442	27	444	103	0.66	1.2	7.2	0.17	1.57	27.6	12.6	2.67	4.21	0.7074	-3.81	1485	-	-	Zhang et al. (2012)
	K-05	KaFS	52.07	2.40	4.90	13.14	3.97	188	40	548	78	0.65	1.3	7.8	0.16	1.28	28.3	11.2	2.71	3.65	-	-	-	-	-	Zhang et al. (2012)
	YG-1	KaFK	46.67	4.18	4.25	15.23	4.78	356	33	579	85	0.74	1.1	8.2	0.13	1.37	32.4	14.1	2.88	3.92	0.7073	-3.78	1477	-	-	Wei et al. (2014)
	YG-2	KaFK	46.58	4.21	4.57	15.40	4.57	349	35	590	83	0.73	1.1	8.2	0.13	1.38	32.3	14.1	2.89	3.92	0.7073	-3.84	1483	-	-	Wei et al. (2014)
	YG-3	KaFK	45.58	4.20	5.35	15.37	4.63	325	38	614	77	0.74	0.9	8.2	0.11	1.29	35.3	15.6	2.79	3.81	0.7065	-2.71	1398	-	-	Wei et al. (2014)
	YG-5	KaFK	46.68	4.02	4.81	14.81	4.90	371	37	564	92	0.72	1.0	8.0	0.13	1.33	32.7	14.1	2.77	3.85	0.7071	-3.57	1470	-	-	Wei et al. (2014)
	YG-6	KaFK	46.56	4.20	4.89	15.76	4.26	354	36	576	84	0.73	1.1	8.1	0.13	1.37	32.0	13.9	2.85	3.92	0.7073	-3.80	1478	-	-	Wei et al. (2014)
	YG-7	KaFK	46.06	3.99	5.05	14.68	3.99	284	38	642	71	0.71	0.9	7.8	0.11	1.28	37.9	14.2	2.64	3.77	0.7062	-2.29	1375	-	-	Wei et al. (2014)
	YG-10	KaFK	48.04	3.84	4.69	14.10	4.07	283	37	614	74	0.64	1.0	6.8	0.15	1.44	26.9	11.8	2.48	3.92	0.7071	-3.66	1500	-	-	Wei et al. (2014)
	YG-11	KaFK	47.76	3.86	5.18	13.97	3.98	278	40	620	72	0.66	1.0	7.1	0.14	1.35	28.4	12.0	2.47	3.87	0.7071	-3.40	1490	-	-	Wei et al. (2014)
	YG-12	KaFK	47.32	3.78	5.46	14.18	3.93	269	41	636	71	0.67	1.0	7.0	0.14	1.34	28.2	12.4	2.43	3.85	0.7070	-3.28	1486	-	-	Wei et al. (2014)
	YG-13	KaFK	47.42	3.76	5.66	13.93	3.84	261	42	633	69	0.59	1.0	6.3	0.16	1.48	25.8	12.5	2.46	3.77	0.7075	-3.31	1491	-	-	Wei et al. (2014)
	YG-14	KaFK	47.08	3.77	5.56	14.03	4.06	262	41	628	69	0.65	0.9	6.9	0.14	1.34	28.7	12.2	2.44	3.80	0.7068	-2.94	1458	-	-	Wei et al. (2014)
	YG-15	KaFK	46.93	3.79	5.86	14.45	4.02	274	42	615	72	0.65	1.0	6.8	0.14	1.35	27.8	13.1	2.41	3.83	0.7067	-2.66	1431	-	-	Wei et al. (2014)
	YG-19	KaFK	44.38	4.99	4.73	16.82	4.32	400	33	601	80	0.75	1.0	8.5	0.11	1.30	34.7	16.5	2.84	3.86	0.7063	-2.73	1391	-	-	Wei et al. (2014)
	YG-20	KaFK	44.51	5.12	4.53	17.00	4.29	410	32	598	80	0.77	0.9	8.7	0.11	1.25	36.1	15.8	2.84	3.82	0.7064	-2.80	1400	-	-	Wei et al. (2014)
	YG-21	KaFK	44.14	4.01	5.02	14.38	7.75	315	38	617	79	0.76	0.9	8.4	0.11	1.27	35.4	15.4	2.74	3.89	0.7066	-2.77	1395	-	-	Wei et al. (2014)

(To be continued)

(Supplementary Table I)

Rock unit	Sample No.	Locality	SiO$_2$ (wt.%)	TiO$_2$ (wt.%)	MgO (wt.%)	FeO$_t$ (wt.%)	Na$_2$O+K$_2$O (wt.%)	Zr (ppm)	Mg#	Ti/Y	Zr/TiO$_2$	Nb/Y	Th/Yb	Nb/Yb	Th/Nb	La/Nb	Nb/U	Ce/Pb	Sm/Yb	La/Sm	^{87}Sr/^{86}Sr$_i$	$\varepsilon_{Nd}(t)$	T_{DM}^{Nd} (Ma)	$\varepsilon_{Hf}(t)$	T_{DM}^{Hf} (Ma)	Data source
Group 1a basalts	YG-22	KaFK	45.34	3.92	5.44	14.83	4.79	294	40	647	75	0.74	0.9	7.9	0.12	1.26	33.8	14.5	2.60	3.85	0.7064	−2.63	1395	-	-	Wei et al. (2014)
	Xhn15-1	Xihenan sec.	46.50	4.37	4.14	15.68	4.61	373	32	580	85	0.69	1.2	7.9	0.15	1.42	28.6	-	2.81	4.01	0.7080	−4.08	1368	-	-	3*
	Xhn16-2	Xihenan sec.	45.83	4.14	4.95	15.31	3.74	348	37	572	84	0.70	1.1	7.5	0.14	1.41	29.4	-	2.71	3.93	0.7076	−3.14	1343	−1.7	1262	1*
	Xhn14-3	Xihenan sec.	46.89	4.36	2.72	12.02	3.36	431	29	582	99	0.82	1.4	9.9	0.14	1.50	28.4	-	3.57	4.16	0.7081	−3.63	1368	−2.4	1199	4*
	GD07-8	Gudongshan sec.	48.14	3.97	4.76	15.47	4.92	293	35	464	74	0.53	1.3	6.4	0.20	1.68	23.6	10.2	2.58	4.19	-	-	-	-	-	Pan et al. (2011)
	GD07-13	Gudongshan sec.	46.43	3.95	4.33	16.96	5.02	413	31	344	105	0.56	1.2	6.7	0.18	1.67	26.8	11.9	2.79	4.00	-	-	-	-	-	Pan et al. (2011)
	GD07-14	Gudongshan sec.	46.46	4.23	4.65	16.11	5.06	418	34	411	99	0.61	1.2	7.4	0.16	1.49	27.5	3.3	2.70	4.09	-	-	-	-	-	Pan et al. (2011)
	GD07-19	Gudongshan sec.	47.17	4.17	5.36	15.30	4.65	327	38	466	78	0.58	1.1	6.9	0.17	1.51	28.5	10.5	2.53	4.11	-	-	-	-	-	Pan et al. (2011)
	Txn25-5	Damusi sec.	45.72	4.50	4.40	14.85	4.32	339	35	636	75	0.72	1.0	7.7	0.13	1.39	32.4	-	2.60	4.08	0.7074	−2.61	1262	0.0	1145	1*
	Txn25-8	Damusi sec.	45.68	4.28	4.48	15.54	4.27	343	34	609	80	0.73	1.1	7.8	0.14	1.37	31.8	13.4	2.77	3.84	-	-	-	-	-	Li et al. (2008)
	Txn25-11	Damusi sec.	45.92	4.28	4.85	15.17	4.14	330	36	631	77	0.73	1.1	8.0	0.13	1.30	32.6	-	2.67	3.93	0.7069	−2.62	1281	-	-	5*
	Txn25-21	Damusi sec.	45.87	4.35	3.86	15.07	4.14	348	31	590	80	0.71	1.0	7.5	0.14	1.36	29.8	12.3	2.59	3.96	0.7069	−2.59	1260	−1.3	1248	1*
	Txn26-3	Damusi sec.	46.86	4.71	3.63	14.21	4.62	403	31	529	86	0.67	1.0	7.4	0.14	1.42	31.4	-	2.71	3.91	-	-	-	-	-	Li et al. (2008)
	Txn26-7	Damusi sec.	46.11	4.45	4.28	14.04	4.11	371	35	603	83	0.74	1.0	8.0	0.13	1.33	24.1	-	2.66	4.12	-	-	-	-	-	Li et al. (2008)
	QMG1102a	Qimugan sec.	50.00	4.64	4.42	14.02	5.02	289	36	738	62	0.64	1.3	6.8	0.19	1.52	18.9	16.4	2.75	3.77	0.7079	−4.65	1500	−2.6	1230	Li et al. (2013)
	QMG1102b	Qimugan sec.	49.99	4.59	4.45	14.20	4.95	287	36	715	62	0.64	1.3	6.9	0.19	1.52	17.6	16.2	2.85	3.69	0.7083	−4.65	1517	−2.5	1229	Li et al. (2013)
	QMG1103a	Qimugan sec.	44.88	4.96	4.26	17.17	5.19	315	31	708	64	0.65	1.3	7.3	0.18	1.47	26.7	12.8	2.70	3.95	-	-	-	-	-	Li et al. (2013)
	QMG1104a	Qimugan sec.	43.82	4.82	5.87	17.92	4.10	313	37	699	65	0.64	1.5	7.0	0.21	1.47	23.7	2.6	2.71	3.81	0.7086	−4.76	1509	−2.4	1223	Li et al. (2013)
	QMG1104b	Qimugan sec.	48.84	3.46	2.16	14.96	5.30	231	20	649	67	0.60	1.5	6.8	0.21	1.70	22.1	13.9	2.85	4.08	-	-	-	-	-	Li et al. (2013)
	QMG1104c	Qimugan sec.	56.38	3.01	1.48	9.73	4.37	201	21	705	67	0.65	1.6	7.2	0.22	1.67	19.9	11.8	2.93	4.10	-	-	-	-	-	Li et al. (2013)
	QMG1105a	Qimugan sec.	45.44	4.30	4.11	15.21	4.53	284	33	674	66	0.67	1.2	7.5	0.16	1.37	28.9	16.9	2.80	3.68	0.7078	−4.19	1445	−2.2	1246	Li et al. (2013)
	QMG1105b	Qimugan sec.	46.96	4.24	4.09	15.05	5.85	279	33	662	66	0.63	1.3	6.9	0.18	1.48	26.5	16.2	2.71	3.76	-	-	-	-	-	Li et al. (2013)
	QMG1105c	Qimugan sec.	45.43	4.30	4.12	15.17	4.54	282	33	689	66	0.66	1.2	7.0	0.17	1.44	26.3	12.6	2.71	3.73	-	-	-	-	-	Li et al. (2013)
	QMG1108a	Qimugan sec.	49.27	4.63	3.88	15.40	4.37	291	31	715	63	0.64	1.3	6.9	0.19	1.42	25.8	10.5	2.71	3.62	0.7078	−4.50	1513	−2.1	1210	Li et al. (2013)
	QMG1108b	Qimugan sec.	51.80	3.45	2.72	8.02	5.91	222	38	639	64	0.65	1.6	7.8	0.20	1.36	23.3	16.7	3.01	3.51	-	-	-	-	-	Li et al. (2013)
	QMG1109a	Qimugan sec.	46.93	4.29	4.82	15.55	5.27	269	36	701	63	0.64	1.2	7.0	0.17	1.45	27.6	16.4	2.62	3.86	0.7079	−3.88	1430	−2.0	1213	Li et al. (2013)

(To be continued)

(Supplementary Table I)

Rock unit	Sample No.	Locality	SiO$_2$ (wt.%)	TiO$_2$ (wt.%)	MgO (wt.%)	FeO$_T$ (wt.%)	Na$_2$O+K$_2$O (wt.%)	Zr (ppm)	Mg#	Ti/Y	Zr/TiO$_2$	Nb/Y	Th/Yb	Nb/Yb	Th/Nb	La/Nb	Nb/U	Ce/Pb	Sm/Yb	La/Sm	^{87}Sr/^{86}Sr$_i$	$\varepsilon_{Nd}(t)$	T_{DM}^{Nd} (Ma)	$\varepsilon_{Hf}(t)$	T_{DM}^{Hf} (Ma)	Data source
Group 1a basalts	QMG1109b	Qimugan sec.	-	-	-	-	-	274	-	-	-	0.62	1.2	6.8	0.17	1.50	26.1	11.2	2.75	3.70	-	-	-	-	-	Li et al. (2013)
	QMG1110a	Qimugan sec.	51.64	4.42	3.74	10.01	5.74	306	40	639	69	0.62	1.3	7.1	0.19	1.63	24.2	11.7	2.94	3.94	-	-	-	-	-	Li et al. (2013)
	QMG1110b	Qimugan sec.	59.46	3.58	2.64	8.44	5.00	220	36	721	61	0.68	1.6	7.4	0.22	1.46	16.7	8.6	2.89	3.73	0.7084	-4.77	1498	-2.8	1238	Li et al. (2013)
	QMG1111b	Qimugan sec.	46.32	4.32	4.30	14.15	4.92	301	35	656	70	0.63	1.3	6.9	0.19	1.59	23.9	16.5	2.75	4.01	0.7079	-4.77	1507	-2.6	1240	Li et al. (2013)
	He4-1-4	He4 bh	47.97	3.97	4.00	15.50	3.29	303	32	591	76	0.63	1.3	6.7	0.19	1.63	25.1	-	2.68	4.09	0.7081	-4.95	1468	0.5	1150	1*
	He4-1-5	He4 bh	47.93	4.01	3.91	15.47	3.40	295	31	612	74	0.63	1.3	7.0	0.19	1.64	24.4	-	2.83	4.05	0.7081	-4.75	1455	-	-	3*
	S99-5	S99 bh	42.94	4.69	3.32	17.14	4.10	355	26	621	76	0.73	1.0	8.1	0.13	1.38	35.3	16.3	3.08	3.62	0.7061	-2.27	1351	-	-	Zhang et al. (2010)
	Z1-1	Z1 bh	49.01	3.67	3.60	13.03	7.06	261	33	611	71	0.67	1.2	6.9	0.17	1.32	27.6	11.3	2.21	4.14	0.7083	-2.59	1407	-	-	Zhang et al. (2010)
	Z1-4	Z1 bh	48.58	3.94	5.10	12.97	6.23	262	41	585	67	0.62	1.1	6.3	0.17	1.39	24.0	10.2	2.34	3.74	-	-	-	-	-	Zhang et al. (2010)
	Ssc21-1	KuFS	50.69	3.07	3.34	14.37	3.98	267	29	504	87	0.68	2.0	7.2	0.28	1.49	14.8	-	2.44	4.43	0.7082	-3.92	1399	-1.3	1253	1*
	Ssc21-5	KuFS	48.16	3.28	4.29	15.05	5.31	286	34	501	87	0.67	2.0	6.9	0.28	1.50	14.6	-	2.29	4.54	-	-3.59	1428	-	-	Yu (2009)
	Ssc21-17	KuFS	49.85	3.19	3.61	15.01	3.62	232	30	567	73	0.68	1.9	7.2	0.26	1.45	16.0	-	2.33	4.49	0.7082	-2.90	1367	-	-	6*
	Ssc22-2	KuFS	48.02	3.83	2.71	16.04	4.38	289	23	589	76	0.72	1.6	7.8	0.21	1.31	21.0	-	2.47	4.11	0.7073	-2.71	1341	-	-	3*
	Ssc22-5	KuFS	46.89	3.69	3.62	16.32	5.14	279	28	583	75	0.73	1.6	7.9	0.20	1.29	19.2	-	2.45	4.19	0.7074	-2.20	1353	-	-	6*
	Ssc22-7	KuFS	46.61	3.65	3.36	16.09	3.36	281	27	567	77	0.70	1.6	7.8	0.21	1.37	18.9	-	2.60	4.08	-	-	-	-	-	Yu (2009)
	Ssc22-11	KuFS	46.70	3.65	3.13	16.17	3.59	284	26	577	78	0.72	1.6	8.0	0.20	1.33	17.0	-	2.41	4.38	-	-1.79	1196	-	-	Yu (2009)
	Ssc22-17	KuFS	45.11	4.06	2.39	14.81	6.02	289	22	601	71	0.71	1.5	7.7	0.19	1.35	15.0	-	2.44	4.26	-	-2.04	1282	-	-	7*
Group 1b basalts	Ssc24-2	KuFS	51.26	2.90	3.35	14.43	3.83	267	29	482	92	0.68	2.0	6.8	0.29	1.58	14.3	-	2.20	4.86	0.7083	-3.36	1311	-2.0	1286	8*
	Ssc24-4	KuFS	51.62	2.92	3.38	13.96	3.84	274	30	470	94	0.66	2.0	7.0	0.29	1.59	14.4	-	2.34	4.73	-	-3.68	1380	-	-	Yu (2009)
	Yg20-5	KuFY	48.83	2.96	3.82	14.82	3.23	221	31	554	75	0.65	1.6	6.9	0.23	1.43	17.4	-	2.37	4.14	-	-	-	-	-	Yu (2009)
	Yg20-7	KuFY	48.28	2.88	3.95	14.80	3.38	223	32	532	77	0.63	1.5	6.6	0.23	1.43	17.5	-	2.25	4.21	-	-	-	-	-	Yu (2009)
	Yg20-13	KuFY	47.36	3.71	3.85	16.24	3.83	287	30	574	77	0.72	1.5	7.8	0.20	1.31	20.2	-	2.47	4.10	-	-	-	-	-	Yu (2009)
	Yg20-15	KuFY	47.31	3.77	3.74	16.41	3.92	293	29	587	78	0.73	1.6	7.9	0.20	1.35	20.4	-	2.55	4.20	-	-	-	-	-	Yu (2009)
	Yg20-16	KuFY	47.56	3.68	3.88	16.36	3.88	295	30	559	80	0.73	1.6	8.1	0.20	1.30	20.0	-	2.57	4.07	-	-	-	-	-	Yu (2009)
	Yg20-18	KuFY	47.40	3.81	3.83	16.53	3.95	295	29	574	77	0.71	1.6	7.9	0.20	1.30	19.9	-	2.57	3.96	-	-	-	-	-	Yu (2009)

(To be continued)

(Supplementary Table I)

Rock unit	Sample No.	Locality	SiO$_2$ (wt.%)	TiO$_2$ (wt.%)	MgO (wt.%)	FeO$_t$ (wt.%)	Na$_2$O+K$_2$O (wt.%)	Zr (ppm)	Mg#	Ti/Y	Zr/TiO$_2$	Nb/Y	Th/Yb	Nb/Yb	Th/Nb	La/Nb	Nb/U	Ce/Pb	Sm/Yb	La/Sm	^{87}Sr/^{86}Sr$_i$	$\varepsilon_{Nd}(t)$	T_{DM}^{Nd} (Ma)	$\varepsilon_{Hf}(t)$	T_{DM}^{Hf} (Ma)	Data source
Group 1b basalts	Yg20-21	KuFY	47.20	3.60	3.91	16.16	4.14	275	30	579	76	0.73	1.6	7.8	0.20	1.27	18.7	-	2.57	3.84	-	-	-	-	-	Yu (2009)
	DWG07-2	KuFK	48.61	3.53	4.15	16.27	4.33	282	31	537	80	0.72	1.8	7.3	0.24	1.42	16.1	9.4	2.60	4.00	-	-	-	-	-	Zhang et al. (2010)
	DWG07-3	KuFK	48.11	3.67	4.17	16.74	4.27	277	31	565	75	0.72	1.6	7.1	0.22	1.39	19.5	8.2	2.51	3.94	-	-	-	-	-	Zhang et al. (2010)
	LKC07-1	KuFK	47.22	3.57	3.88	17.04	4.56	282	29	535	79	0.71	1.6	7.1	0.23	1.40	18.6	8.4	2.23	4.46	0.7072	-2.34	1402	-	-	Zhang et al. (2010)
	LKC07-2	KuFK	47.84	3.82	4.21	16.88	4.37	279	31	573	73	0.70	1.7	7.6	0.23	1.40	19.3	8.6	2.65	4.01	-	-	-	-	-	Zhang et al. (2010)
	LKC07-3	KuFK	51.93	2.86	3.33	13.80	4.79	247	30	471	86	0.67	2.2	7.1	0.31	1.59	13.9	4.4	2.45	4.57	-	-	-	-	-	Zhang et al. (2010)
	Yg01	KuFY	51.40	2.96	3.14	13.73	4.82	250	29	534	85	0.73	2.1	7.5	0.28	1.55	14.7	7.4	2.49	4.64	0.7080	-4.13	1424	-	-	Li Y.-Q. et al. (2012)
	Yg04	KuFY	46.70	3.82	3.96	16.59	4.80	283	30	623	74	0.79	1.8	8.9	0.21	1.30	19.7	11.5	2.74	4.23	0.7072	-2.05	1277	-	-	Li Y.-Q. et al. (2012)
	P1-1	KuFS	50.39	3.09	4.36	14.41	4.24	224	35	542	72	0.69	1.7	7.4	0.22	1.41	16.7	4.7	2.51	4.17	0.7081	-3.63	1526	-	-	Yu et al. (2012)
	P1-2	KuFS	50.22	3.13	4.46	14.30	4.13	226	36	559	72	0.70	1.6	7.6	0.21	1.39	17.3	4.0	2.51	4.22	0.7081	-3.68	1511	-	-	Yu et al. (2012)
	P1-3	KuFS	50.37	3.07	4.39	14.34	4.14	229	35	541	75	0.70	1.6	7.6	0.21	1.39	17.2	3.1	2.53	4.16	0.7081	-3.82	1520	-	-	Yu et al. (2012)
	P1-4	KuFS	50.52	3.08	4.48	14.24	4.21	233	36	529	76	0.69	1.6	7.5	0.21	1.40	17.0	4.8	2.42	4.35	0.7080	-3.50	1427	-	-	Yu et al. (2012)
	P1-5	KuFS	52.60	2.93	3.65	13.25	4.66	252	33	476	86	0.68	2.0	7.7	0.26	1.47	15.1	2.6	2.51	4.50	0.7083	-4.14	1463	-	-	Yu et al. (2012)
	P2-8	KuFS	46.04	3.93	3.93	17.09	5.67	298	29	568	76	0.75	1.4	8.4	0.16	1.05	20.1	5.2	2.47	3.58	0.7080	-2.90	1500	-	-	Yu et al. (2012)
	P2-10	KuFS	48.43	3.66	3.92	15.88	4.77	278	31	574	76	0.75	1.5	8.4	0.17	1.25	20.5	5.2	2.57	4.08	0.7074	-2.71	1396	-	-	Yu et al. (2012)
	P2-11	KuFS	48.23	3.64	4.17	15.91	4.85	278	32	562	76	0.73	1.5	8.5	0.17	1.24	20.2	4.7	2.53	4.17	0.7072	-2.47	1320	-	-	Yu et al. (2012)
	P2-12	KuFS	48.06	3.59	4.46	16.31	4.68	277	33	548	77	0.73	1.4	8.5	0.17	1.23	21.0	7.3	2.51	4.19	0.7080	-2.85	1395	-	-	Yu et al. (2012)
	KP-01	KuFS	48.53	3.95	3.56	17.02	4.31	290	27	591	73	0.72	1.6	7.7	0.21	1.39	18.9	7.8	2.61	4.07	-	-	-	-	-	Zhang et al. (2012)
	KP-02	KuFS	48.87	4.04	3.25	15.41	6.34	308	27	580	76	0.74	1.6	7.7	0.21	1.35	18.1	8.1	2.31	4.51	0.7077	-2.22	1390	-	-	Zhang et al. (2012)
	KP-03	KuFS	48.11	3.99	3.71	17.01	6.74	314	28	557	79	0.74	1.7	7.8	0.22	1.38	19.8	7.8	2.65	4.04	-	-	-	-	-	Zhang et al. (2012)
	KP-04	KuFS	61.24	3.05	2.47	15.30	4.07	214	22	686	70	0.83	2.1	9.5	0.23	1.37	8.4	9.7	2.82	4.63	0.7085	-2.63	1403	-	-	Zhang et al. (2012)
	KP-05	KuFS	46.97	3.78	4.00	16.91	4.12	293	30	601	78	0.79	1.9	9.1	0.21	1.27	22.8	8.0	2.83	4.08	-	-	-	-	-	Zhang et al. (2012)
	KP-06	KuFS	49.69	3.34	3.94	15.32	4.16	242	31	584	72	0.72	1.9	7.6	0.26	1.47	19.0	8.5	2.53	4.41	-	-	-	-	-	Zhang et al. (2012)
	KP-09	KuFS	48.48	4.23	3.84	16.92	3.99	291	29	628	69	0.71	1.7	7.8	0.21	1.39	19.2	8.6	2.43	4.46	0.7085	-2.54	1419	-	-	Zhang et al. (2012)
	KP-10	KuFS	56.44	3.67	2.07	13.27	6.61	269	22	694	73	0.85	1.9	9.0	0.22	1.25	13.4	11.2	2.83	3.97	-	-	-	-	-	Zhang et al. (2012)
	H1-6-1	H1 bh	49.13	2.95	3.09	15.16	5.91	330	27	412	112	0.59	1.5	6.3	0.24	1.42	11.4	-	2.20	4.06	0.7075	-3.24	1358	-0.3	1183	1*

(To be continued)

(Supplementary Table I)

Rock unit	Sample No.	Locality	SiO$_2$ (wt.%)	TiO$_2$ (wt.%)	MgO (wt.%)	FeO$_t$ (wt.%)	Na$_2$O+K$_2$O (wt.%)	Zr (ppm)	Mg#	Ti/Y	Zr/TiO$_2$	Nb/Y	Th/Yb	Nb/Yb	Th/Nb	La/Nb	Nb/U	Ce/Pb	Sm/Yb	La/Sm	^{87}Sr/^{86}Sr$_i$	$\varepsilon_{Nd}(t)$	T_{DM}^{Nd} (Ma)	$\varepsilon_{Hf}(t)$	T_{DM}^{Hf} (Ma)	Data source
Group 1b basalts	H1-6-3	H1 bh	51.01	2.84	3.34	13.96	4.60	325	30	390	114	0.56	1.4	5.9	0.23	1.46	17.3	-	2.18	3.95	0.7075	-3.25	1391	-	-	Yu (2009)
	H1-6-6	H1 bh	49.59	2.88	3.39	14.05	4.85	331	30	384	115	0.56	1.3	5.7	0.24	1.43	17.1	-	2.11	3.86	0.7053	-1.02	1070	0.8	983	3*
	SL1-8-10	SL1 bh	49.05	2.50	7.54	10.95	4.09	221	55	685	89	1.39	3.0	17.9	0.17	1.20	29.1	12.5	4.17	5.16	0.7053	-	-	-	-	1*
	SL1-8-11	SL1 bh	48.43	2.43	8.31	11.44	4.14	215	56	689	89	1.40	3.0	18.0	0.17	1.16	30.2	10.4	4.17	5.03	-	-	-	-	-	Yu (2009)
	SL1-8-14	SL1 bh	54.55	1.98	6.55	9.82	4.80	298	54	448	150	1.54	3.1	18.1	0.17	1.07	29.3	-	3.67	5.31	0.7053	-1.79	1139	-	-	3*
	YM5-22-8	YM5 bh	51.69	2.75	5.87	11.06	4.52	245	49	691	89	1.38	3.1	17.2	0.18	1.18	24.3	15.6	4.00	5.09	0.7050	-1.15	1100	0.9	989	1*
	YM5-23-7	YM5 bh	52.29	2.56	6.15	11.93	1.78	228	48	676	89	1.37	3.0	17.0	0.18	1.13	26.5	-	3.97	4.83	-	-	-	-	-	Yu (2009)
	YM5-23-10	YM5 bh	51.63	2.57	5.26	11.35	4.35	227	45	674	88	1.34	3.2	17.2	0.18	1.17	25.3	-	4.08	4.96	0.7052	-1.91	1176	-	-	3*
	YM5-23-17	YM5 bh	50.24	2.65	7.34	11.81	3.42	192	53	768	73	1.34	2.5	17.3	0.14	1.12	34.0	-	3.98	4.88	-	-	-	-	-	Yu (2009)
	YM8-13-7	YM8 bh	53.46	2.77	3.75	9.29	5.89	290	42	722	105	1.67	3.7	19.5	0.19	1.38	22.6	-	4.31	6.22	0.7052	-1.13	1032	0.5	980	1*
	YM8-13-29	YM8 bh	54.94	2.27	3.69	11.57	4.96	249	36	626	110	1.50	3.7	19.2	0.19	1.36	24.2	-	4.18	6.26	0.7053	-1.75	1083	-	-	3*
	YT6-9-55	YT6 bh	47.39	3.22	7.80	13.28	5.51	251	51	669	78	1.26	2.3	15.9	0.14	1.04	33.7	-	3.61	4.57	0.7050	0.08	1047	1.4	975	1*
	YT6-9-66	YT6 bh	49.77	2.92	5.22	13.44	6.56	229	41	660	78	1.25	2.2	15.6	0.14	1.09	60.3	-	3.75	4.56	-	-	-	-	-	Yu (2009)
	YT6-9-80	YT6 bh	50.49	2.59	5.62	11.72	4.02	208	46	629	80	1.20	2.1	15.0	0.14	1.06	26.4	-	3.60	4.42	0.7050	0.22	1042	-	-	3*
Group 2 basalts	SL1-2	SL1 bh	53.52	2.07	3.81	10.29	3.85	233	40	474	113	1.45	1.6	16.9	0.09	0.79	47.5	12.5	3.21	4.18	0.7051	-2.54	1381	-	-	Tian et al. (2010)
	SL1-3	SL1 bh	48.83	2.30	5.16	11.03	4.28	180	45	573	78	1.17	1.4	15.3	0.09	0.89	51.4	10.4	3.40	4.01	0.7050	-1.30	1252	-	-	Tian et al. (2010)
	SL1-4	SL1 bh	49.29	2.52	7.65	10.11	4.03	162	57	818	64	1.51	2.5	21.3	0.12	1.04	47.0	15.6	4.33	5.11	-	-	-	-	-	Tian et al. (2010)
	YM5-1	YM5 bh	51.30	2.62	8.79	10.90	2.88	176	59	778	67	1.32	2.3	17.2	0.13	1.09	34.0	24.5	3.92	4.81	0.7052	-2.07	1332	-	-	Tian et al. (2010)
	YM5-4	YM5 bh	48.92	2.93	9.07	12.37	3.03	203	57	782	69	1.41	2.9	19.7	0.15	1.08	30.7	23.7	4.22	5.02	0.7050	-1.30	1252	-	-	Tian et al. (2010)
	YM5-5	YM5 bh	48.34	2.84	8.19	11.26	3.59	200	56	775	70	1.40	2.1	18.7	0.11	1.10	38.0	23.2	3.87	5.32	0.7052	-2.07	1332	-	-	Tian et al. (2010)
	YM5-6	YM5 bh	48.20	3.09	6.86	11.66	4.07	206	51	861	66	1.47	1.9	17.0	0.11	1.05	30.9	24.4	3.67	4.88	0.7049	-1.62	1297	-	-	Tian et al. (2010)
	YM5-9	YM5 bh	49.04	2.76	9.03	12.17	2.58	185	57	793	67	1.38	2.4	18.4	0.13	1.10	37.3	22.7	4.04	4.99	0.7050	-1.77	1303	-	-	Tian et al. (2010)
	YM8-4	YM8 bh	47.72	3.02	7.88	11.05	3.02	140	56	1331	46	1.95	2.5	19.8	0.13	0.88	50.1	29.9	4.01	4.34	-	-	-	-	-	Tian et al. (2010)
	YT6-1	YT6 bh	47.44	3.14	5.97	12.43	4.24	168	46	1102	54	1.68	2.4	16.2	0.15	0.97	29.8	16.2	3.24	4.85	-	-	-	-	-	Tian et al. (2010)
	YT6-2	YT6 bh	50.51	3.01	4.70	11.69	7.18	167	42	1186	56	1.87	2.4	16.8	0.14	0.89	63.5	16.8	3.30	4.53	-	-	-	-	-	Tian et al. (2010)

(To be continued)

(Supplementary Table I)

Rock unit	Sample No.	Locality	SiO₂ (wt.%)	TiO₂ (wt.%)	MgO (wt.%)	FeO_T (wt.%)	Na₂O+K₂O (wt.%)	Zr (ppm)	Mg#	Ti/Y	Zr/TiO₂	Nb/Y	Th/Yb	Nb/Yb	Th/Nb	La/Nb	Nb/U	Ce/Pb	Sm/Yb	La/Sm	$^{87}Sr/^{86}Sr_i$	$\varepsilon_{Nd}(t)$	T_{DM}^{Nd} (Ma)	$\varepsilon_{Hf}(t)$	T_{DM}^{Hf} (Ma)	Data source
Ultramafic–Mafic intrusive rocks																										
	BC1116	Wajilitag	45.77	3.61	12.36	12.34	1.57	151	64	1410	42	1.21	1.5	16.0	0.09	0.75	41.3	-	4.88	2.46	0.7042	1.62	1231	3.0	849	Cao et al. (2014)
	BC1119	Wajilitag	46.17	3.58	12.23	12.57	1.60	153	63	1489	43	1.27	1.7	16.3	0.11	0.77	33.2	-	4.88	2.57	0.7042	1.85	1164	-	-	Cao et al. (2014)
	BC1123	Wajilitag	32.73	8.54	10.88	25.41	0.26	94	43	3800	11	0.72	0.1	9.4	0.01	0.41	483.0	-	4.75	0.80	0.7044	2.03	2005	1.8	914	Cao et al. (2014)
	BC1124	Wajilitag	32.75	8.51	11.15	24.95	0.35	89	44	3778	10	0.93	0.1	13.0	0.01	0.30	420.0	-	4.85	0.81	0.7046	1.87	1996	-	-	Cao et al. (2014)
	BC1126	Wajilitag	34.39	7.68	11.36	23.09	0.37	91	47	3332	12	0.73	0.1	10.5	0.01	0.40	337.0	-	5.97	0.70	0.7044	2.34	1264	0.3	991	Cao et al. (2014)
	BC1128	Wajilitag	33.18	8.27	10.93	24.78	0.31	105	44	3253	13	0.84	0.1	12.0	0.01	0.34	426.3	-	4.90	0.82	0.7044	2.54	2127	-	-	Cao et al. (2014)
	BC1131	Wajilitag	34.53	7.56	11.44	23.10	0.42	94	47	3199	12	0.72	0.3	10.5	0.03	0.42	203.6	-	5.23	0.84	0.7045	2.47	1405	1.5	930	Cao et al. (2014)
	BC1140	Wajilitag	34.66	7.70	11.51	23.12	0.28	93	47	3250	12	0.77	0.1	10.7	0.01	0.36	544.5	-	4.71	0.82	0.7045	2.43	1876	-0.3	1010	Cao et al. (2014)
Clinopyroxenite	BC1146	Wajilitag	33.29	8.65	10.97	24.10	0.38	94	45	4337	11	0.80	0.4	10.8	0.04	0.41	7.3	-	4.75	0.94	0.7050	1.62	2296	0.3	965	Cao et al. (2014)
	BC1141	Wajilitag	36.36	7.25	10.92	21.86	0.60	108	47	2471	15	2.16	0.3	32.2	0.01	0.22	542.3	-	8.53	0.85	0.7042	2.81	1405	-	-	Cao et al. (2014)
	BC1148	Wajilitag	36.67	7.21	11.07	21.93	0.36	163	47	2406	23	1.37	0.3	19.5	0.02	0.78	66.4	-	5.68	2.69	0.7051	3.96	873	3.0	848	Cao et al. (2014)
	BC1171	Wajilitag	42.50	3.67	14.04	15.34	1.73	193	62	1056	53	1.22	1.5	17.3	0.09	1.36	44.7	-	6.18	3.83	0.7038	2.49	879	4.0	813	Cao et al. (2014)
	BC1172	Wajilitag	44.82	4.04	7.44	13.64	4.04	279	49	795	69	1.87	2.0	25.8	0.08	1.05	49.4	-	6.40	4.22	0.7047	1.17	946	-	-	Cao et al. (2014)
	BC1173	Wajilitag	41.72	4.98	11.07	15.83	1.63	169	55	1698	34	1.45	1.2	19.7	0.06	0.79	65.1	-	5.59	2.79	0.7037	2.68	981	2.7	866	Cao et al. (2014)
	BC1175	Wajilitag	41.38	5.15	11.18	16.43	1.44	169	55	1802	33	1.49	1.2	20.6	0.06	0.71	64.0	-	5.56	2.64	0.7038	2.35	1044	-	-	Cao et al. (2014)
	Wjl6-1	Wajilitag	37.89	5.40	11.15	18.99	0.70	159	51	1965	29	0.57	1.0	9.0	0.11	1.18	31.4	-	6.11	1.73	0.7050	1.90	1241	3.6	802	1*
	Wjl6-7	Wajilitag	37.54	5.38	9.87	19.77	1.06	196	47	1837	36	0.24	2.4	3.8	0.64	5.14	8.78	-	6.59	2.95	0.7051	1.62	1105	-	-	3*
	05BH-12	Wajilitag	51.42	2.38	2.69	13.97	6.45	205	26	227	86	1.08	1.1	17.4	0.06	1.41	67.4	-	6.25	3.92	0.7567	0.55	1015	-	-	Zhang et al. (2008)
Gabbro	BC1102	Wajilitag	44.36	4.94	6.03	14.68	3.78	258	42	1078	52	1.99	2.4	25.3	0.09	0.85	42.4	-	5.32	4.03	0.7043	1.83	916	-	-	Cao et al. (2014)
	BC1103	Wajilitag	43.44	5.41	7.96	14.70	3.79	359	49	1034	66	2.15	2.2	26.3	0.08	0.69	49.3	-	5.02	3.59	0.7043	1.09	1003	3.0	853	Cao et al. (2014)
	BC1104	Wajilitag	44.01	4.67	7.18	14.54	3.74	267	47	1102	57	1.74	2.2	22.6	0.10	0.89	35.2	-	5.44	3.70	0.7044	1.04	1016	3.4	834	Cao et al. (2014)
Ultramafic dike	X01-5a	Xiaohaizi	43.03	1.60	22.55	15.95	1.02	97	72	808	60	1.15	1.5	13.9	0.11	0.98	35.3	13.7	3.70	3.68	0.7042	1.98	1018	-	-	Yu (2009)
	X01-5b	Xiaohaizi	44.03	1.72	20.76	15.19	0.98	95	71	856	55	1.10	1.5	13.1	0.12	0.95	32.0	2.1	3.57	3.50	0.7044	3.17	955	-	-	Yu (2009)

(To be continued)

(Supplementary Table 1)

Rock unit	Sample No.	Locality	SiO_2 (wt.%)	TiO_2 (wt.%)	MgO (wt.%)	FeO_t (wt.%)	Na_2O+K_2O (wt.%)	Zr (ppm)	Mg#	Ti/Y	Zr/TiO_2	Nb/Y	Th/Y	Nb/Yb	Th/Nb	La/Nb	Nb/U	Ce/Pb	Sm/Yb	La/Sm	$^{87}Sr/^{86}Sr_i$	$\varepsilon_{Nd}(t)$	T_{DM}^{Nd} (Ma)	$\varepsilon_{Hf}(t)$	T_{DM}^{Hf} (Ma)	Data source
Ultramafic dike	Xhzn2-1	Xiaohaizi	43.42	1.64	20.43	16.14	0.43	91	69	839	56	1.06	1.4	12.2	0.11	0.99	27.7	-	3.57	3.38	0.7043	2.49	983	4.9	810	1*
	Xhzn2-3	Xiaohaizi	41.84	1.72	22.28	17.84	0.42	86	69	953	50	1.12	1.5	12.7	0.12	1.04	35.4	-	3.61	3.65	0.7040	2.35	981	4.5	831	1*
	XHZ62	Xiaohaizi	44.84	1.66	19.54	13.68	1.03	110	72	681	66	1.00	1.3	14.7	0.09	0.84	42.9	22.5	3.71	3.32	0.7048	2.91	974	-	-	Zhou et al. (2009)
	Wjl7-1	Wajilitag	45.77	3.09	3.45	10.30	10.73	612	37	406	198	2.29	4.1	34.4	0.12	0.91	32.7	-	5.70	5.49	0.7036	3.79	712	4.7	764	1*
	Wjl7-4	Wajilitag	45.67	3.29	3.51	11.00	9.93	574	36	418	174	2.10	3.7	31.7	0.12	0.93	32.1	-	5.57	5.27	0.7036	3.54	734	-	-	3*
	S03-4a	Xiaohaizi	46.97	4.28	5.02	13.73	5.54	295	39	843	69	1.70	2.1	21.5	0.10	0.84	39.4	3.2	4.20	4.31	0.7038	3.64	764	-	-	Yu (2009)
	S03-6a	Xiaohaizi	46.53	3.61	6.10	13.02	6.02	257	46	857	71	1.52	2.1	18.3	0.11	0.93	33.3	13.0	4.15	4.12	0.7044	3.55	832	-	-	Yu (2009)
	X01-3a	Xiaohaizi	48.82	2.86	6.18	12.13	4.82	228	48	754	80	1.39	2.3	15.1	0.15	1.17	25.3	11.25	3.84	4.59	0.7053	-1.07	1194	-	-	Yu (2009)
	X-4	Xiaohaizi	47.16	3.92	5.02	15.35	3.02	294	37	731	75	1.40	2.0	20.0	0.10	0.83	34.2	-	4.09	4.06	-	4.54	716	-	-	Yu (2009)
	X10-1	Xiaohaizi	52.30	2.76	3.55	11.12	5.64	414	36	421	150	0.53	2.1	6.7	0.31	3.12	9.6	-	4.88	4.30	-	5.23	619	-	-	Yu (2009)
	X10-4	Xiaohaizi	52.03	2.87	3.92	10.85	5.81	459	39	391	160	1.82	2.5	25.5	0.10	0.82	38.0	-	4.84	4.31	-	5.16	610	-	-	Yu (2009)
	X11-3	Xiaohaizi	45.84	3.76	5.06	13.42	4.75	269	40	749	71	1.41	1.6	18.4	0.08	0.91	41.7	-	3.73	3.48	0.7037	2.67	844	-	-	3*
	Xhzn3-1	Xiaohaizi	52.96	3.06	3.78	11.53	4.35	325	37	554	106	1.19	2.9	13.4	0.22	1.38	21.2	-	3.40	5.44	0.7061	-2.53	1229	-0.9	1079	8*
	Xhzn3-3	Xiaohaizi	52.01	2.76	3.55	10.85	5.63	309	37	519	112	1.30	2.2	14.2	0.16	1.23	25.8	-	3.56	4.89	0.7058	-0.76	1159	-	-	3*
Mafic dike	Xhzn4-12	Xiaohaizi	47.09	3.83	5.04	14.21	3.68	276	39	793	72	1.30	2.4	17.1	0.14	1.00	29.9	-	4.19	4.08	0.7042	2.82	894	3.1	876	9*
	XHZ36	Xiaohaizi	48.70	3.27	4.64	12.33	4.82	362	40	504	111	1.59	1.2	20.5	0.06	0.79	24.0	2.2	3.84	4.21	0.7048	5.24	643	-	-	Zhou et al. (2009)
	XHZ54	Xiaohaizi	48.05	3.41	5.22	13.15	5.63	320	41	666	94	1.43	1.5	19.6	0.08	0.85	31.2	30.8	4.55	3.65	0.7051	1.72	1015	-	-	Zhou et al. (2009)
	XHZ07-7	Xiaohaizi	49.32	3.61	4.22	12.38	4.93	413	38	578	114	1.76	3.2	24.5	0.13	0.97	31.2	6.5	5.01	4.75	0.7048	4.69	705	-	-	Zhang et al. (2010)
	BC-0	Xiaohaizi	52.04	3.08	3.67	11.11	5.22	305	37	594	99	1.46	2.3	18.3	0.12	1.02	33.8	19.7	3.94	4.74	0.7057	0.01	1099	-	-	Wei et al. (2014)
	BC-2	Xiaohaizi	46.84	3.86	4.50	13.94	4.54	255	37	848	66	1.76	1.9	23.1	0.08	0.73	43.7	20.2	4.06	4.15	-	4.73	722	-	-	Wei et al. (2014)
	BC-3	Xiaohaizi	50.06	3.41	3.93	12.34	4.95	275	36	730	81	1.50	2.2	18.8	0.12	0.92	34.9	19.0	3.88	4.45	0.7068	0.60	1079	-	-	Wei et al. (2014)
	BC-4	Xiaohaizi	45.52	4.07	5.35	15.44	3.34	218	38	999	54	1.53	1.7	20.3	0.09	0.76	42.4	16.3	3.98	3.85	0.7057	4.54	767	-	-	Wei et al. (2014)
	BC-5	Xiaohaizi	46.11	4.70	5.01	14.19	4.81	299	39	969	64	1.97	2.2	26.5	0.08	0.76	44.1	25.0	4.54	4.43	0.7063	4.47	719	-	-	Wei et al. (2014)
	BC-6	Xiaohaizi	47.08	4.08	5.29	14.81	2.99	233	39	926	57	1.40	1.7	18.0	0.10	0.83	39.3	19.0	3.82	3.90	0.7047	4.08	805	-	-	Wei et al. (2014)
	BC-9	Xiaohaizi	50.53	3.49	3.94	12.13	4.37	318	37	637	91	1.39	2.8	19.9	0.14	1.04	33.5	16.3	4.34	4.75	0.7058	-0.26	1079	-	-	Wei et al. (2014)
	BC-12	Xiaohaizi	50.38	3.21	3.60	10.70	6.19	298	38	592	93	1.42	1.8	17.6	0.10	0.97	38.3	22.8	3.83	4.44	0.7047	1.05	1025	-	-	Wei et al. (2014)

(To be continued)

(Supplementary Table 1)

Rock unit	Sample No.	Locality	SiO_2 (wt.%)	TiO_2 (wt.%)	MgO (wt.%)	FeO_T (wt.%)	Na_2O+K_2O (wt.%)	Zr (ppm)	Mg#	Ti/Y	Zr/TiO_2	Nb/Y	Th/Yb	Nb/Yb	Th/Nb	La/Nb	Nb/U	Ce/Pb	Sm/Yb	La/Sm	$^{87}Sr/^{86}Sr_i$	$\varepsilon_{Nd}(t)$	T_{DM}^{Nd} (Ma)	$\varepsilon_{Hf}(t)$	T_{DM}^{Hf} (Ma)	Data source
	Yg17-2	Yingan sec.	47.54	3.03	5.13	15.21	3.07	221	38	558	73	0.59	0.9	7.4	0.12	1.07	33.3	-	2.57	3.09	-	3.56	975	-	-	Yu (2009)
	Yg17-5	Yingan sec.	47.86	3.12	5.08	14.50	3.07	219	38	596	70	0.63	1.0	7.9	0.12	1.02	32.9	-	2.52	3.20	-	3.98	816	-	-	Yu (2009)
	Yg18-4	Yingan sec.	48.94	2.74	4.51	13.14	3.23	168	38	649	61	0.66	1.1	7.9	0.14	1.20	21.2	-	2.69	3.52	-	3.63	851	-	-	Yu (2009)
	Yg19-1	Yingan sec.	52.22	2.41	4.20	13.43	3.81	188	36	552	78	0.59	1.1	7.0	0.16	1.26	23.3	-	2.43	3.64	-	2.13	987	-	-	Yu (2009)
	Yg19-3	Yingan sec.	51.62	2.28	4.63	13.94	3.56	183	37	510	80	0.59	1.1	7.2	0.16	1.24	25.4	-	2.26	3.96	-	1.81	1075	-	-	Yu (2009)
	Yg19-4	Yingan sec.	52.09	2.31	3.87	13.45	3.80	175	34	568	76	0.62	1.2	7.6	0.15	1.25	24.5	-	2.78	3.41	-	1.93	1016	-	-	Yu (2009)
	DWG07-1	Keping area	48.21	3.71	4.14	16.25	4.80	284	31	557	76	0.71	1.7	7.4	0.23	1.40	17.2	8.6	2.33	4.46	0.7074	-2.13	1374	-	-	Zhang et al. (2010)
Mafic dike	DWG07-4	Keping area	48.30	2.96	5.66	15.76	3.49	201	39	575	68	0.64	0.9	6.9	0.13	1.13	32.9	11.6	2.53	3.07	0.7048	3.43	1024	-	-	Zhang et al. (2010)
	TWC07-1	Keping area	45.10	2.46	8.74	14.52	3.70	165	52	621	67	0.66	1.0	6.6	0.15	1.40	26.6	8.85	2.38	3.87	0.7065	-2.66	1430	-	-	Zhang et al. (2010)
	Z1-6	Z1 bh	45.99	3.44	5.57	14.51	4.26	238	41	569	69	0.62	1.1	6.2	0.18	1.42	24.3	11.3	2.08	4.20	0.7069	-2.57	1426	-	-	Zhang et al. (2010)
	Z16-2	Z16 bh	44.52	3.89	5.55	15.83	4.41	262	38	606	67	0.65	1.1	6.6	0.17	1.39	24.9	19.7	2.21	4.13	0.7071	-2.77	1439	-	-	Zhang et al. (2010)
	Z16-6	Z16 bh	43.29	3.69	4.97	15.63	4.56	248	36	591	67	0.66	1.1	6.6	0.17	1.43	20.6	17.2	2.17	4.35	0.7076	-2.22	1371	-	-	Zhang et al. (2010)
	DH12-1	DH12 bh	44.51	3.01	12.34	12.92	4.34	183	63	787	61	1.72	2.1	22.1	0.10	0.74	41.6	20.4	3.40	4.83	0.7060	-1.06	1351	-	-	Tian et al. (2010)
	DH12-2	DH12 bh	44.86	2.57	8.84	14.13	4.87	165	53	750	64	1.79	2.0	23.2	0.09	0.82	34.7	18.1	3.90	4.86	0.7054	-0.36	1228	-	-	Tian et al. (2010)
	YM201-1	YM201 bh	49.20	2.98	5.91	11.75	3.38	205	47	794	69	1.55	2.3	20.8	0.11	0.89	43.5	16.4	4.65	3.99	0.7065	-4.05	1602	-	-	Tian et al. (2010)
Syenitic rocks																										
	05BH-1	Wajilitag	62.22	0.55	0.60	3.23	10.69	337	25	163	616	2.38	4.9	31.7	0.16	1.26	27.8	-	4.44	8.97	0.7560	2.59	737	-	-	Zhang et al. (2008)
	05BH-4	Wajilitag	64.24	0.42	0.36	3.31	11.63	614	16	75	1456	2.04	5.5	26.3	0.21	1.20	21.9	-	4.29	7.34	0.7224	2.99	719	-	-	Zhang et al. (2008)
	05BH-6	Wajilitag	60.85	0.75	0.89	4.67	10.55	466	25	136	619	1.84	2.4	25.0	0.10	1.20	35.1	-	4.88	6.18	0.7036	2.45	779	-	-	Zhang et al. (2008)
	05BH-7	Wajilitag	62.79	0.56	0.58	3.75	10.95	591	22	135	1047	2.08	4.3	25.5	0.17	1.06	26.9	-	4.30	6.27	0.7030	2.51	798	-	-	Zhang et al. (2008)
Syenite	05BH-9	Wajilitag	59.97	0.88	1.16	5.15	9.57	323	29	151	366	2.15	3.3	31.6	0.10	1.28	26.3	-	5.88	6.85	0.7035	2.58	763	-	-	Zhang et al. (2008)
	05BH-10	Wajilitag	56.67	1.47	2.19	7.25	9.30	297	35	287	202	1.73	3.7	23.0	0.16	1.23	33.7	-	4.76	5.95	0.7031	1.52	862	-	-	Zhang et al. (2008)
	Xhz0506-1a	Xiaohaizi	62.95	0.62	0.57	3.11	11.29	503	24	248	815	2.51	3.2	27.3	0.12	0.99	32.5	-	3.53	7.63	0.7041	3.43	698	-	-	Yu (2009)
	Xhz0506-8	Xiaohaizi	62.55	0.63	0.67	3.39	10.83	675	26	175	1075	2.49	4.4	27.1	0.16	1.32	29.5	-	4.50	7.93	0.7036	3.74	656	-	-	Yu (2009)
Quartz syenite	05BH-3	Wajilitag	68.71	0.11	0.01	1.85	12.00	564	1	16	5110	2.06	5.3	28.1	0.19	0.85	21.3	-	3.69	6.45	0.6667	1.56	899	-	-	Zhang et al. (2008)

(To be continued)

(Supplementary Table I)

Rock unit	Sample No.	Locality	SiO_2 (wt.%)	TiO_2 (wt.%)	MgO (wt.%)	FeO_T (wt.%)	Na_2O+K_2O (wt.%)	Zr (ppm)	Mg#	Ti/Y	Zr/TiO_2	Nb/Y	Th/Yb	Nb/Yb	Th/Nb	La/Nb	Nb/U	Ce/Pb	Sm/Yb	La/Sm	$^{87}Sr/^{86}Sr_i$	$\varepsilon_{Nd}(t)$	T_{DM}^{Nd} (Ma)	$\varepsilon_{Hf}(t)$	T_{DM}^{Hf} (Ma)	Data source
Quartz syenite	05BH-5	Wajilitag	68.18	0.34	0.16	4.64	10.25	1349	6	28	3949	2.88	4.7	31.3	0.15	0.85	16.9	-	3.52	7.54	0.6844	2.29	775	-	-	Zhang et al. (2008)
Syenitic porphyry	S03-1b	Xiaohaizi	67.56	0.31	0.57	3.08	10.95	897	25	35	2863	2.80	4.7	30.3	0.15	0.79	24.1	28.0	3.69	6.49	0.7103	3.35	730	-	-	Yu (2009)
	Xhzn4-4	Xiaohaizi	67.94	0.30	0.38	3.61	10.32	1007	16	28	3312	2.49	4.7	27.7	0.17	0.90	25.1	-	3.61	6.92	-	4.12	669	5.3	732	10*
	Xhzn4-7	Xiaohaizi	67.09	0.30	0.32	4.70	10.51	1000	11	30	3303	2.60	4.7	30.1	0.16	0.86	23.8	-	3.68	7.00	0.6995	3.96	679	-	-	Yu (2009)
	Xhzn4-8	Xiaohaizi	68.22	0.30	0.46	3.88	10.51	920	18	33	3037	2.73	4.9	29.5	0.17	0.83	23.3	-	3.52	7.00	0.7035	4.05	657	7.1	654	11*
Picrites	YT1-1	YT1 bh	45.60	2.01	18.11	13.16	1.64	99	71	1074	50	1.44	2.1	15.4	0.14	0.90	35.0	20.0	4.21	3.28	-	-	-	-	-	Tian et al. (2010)
	YT1-2	YT1 bh	45.00	2.00	17.68	13.55	1.84	99	70	1070	49	1.41	1.8	13.8	0.13	0.91	35.0	16.3	3.67	3.41	-	-	-	-	-	Tian et al. (2010)
	YH5-1	YH5 bh	46.69	2.47	15.67	12.12	2.75	192	70	748	78	2.02	3.5	27.6	0.13	1.05	39.0	30.3	5.03	5.74	0.7070	-5.18	1528	-	-	Tian et al. (2010)

Note: Major elements are normalized to volatile-free compositions, and FeO_T is the total FeO; $^{87}Sr/^{86}Sr_i$, $\varepsilon_{Nd}(t)$ and $\varepsilon_{Hf}(t)$ are recalculated at T=290 Ma for the basalts and picrites, T=283 Ma for the ultramafic–mafic intrusive rocks and T=274 Ma for the syenitic rocks; T_{DM}^{Nd} and TDMHf are the Nd- and Hf-depleted mantle model ages, respectively; the related localities are shown in Fig. 3.1 in the book. KaFK=Kaipaizileike Formation, Keping area; KaFY=Kaipaizileike Formation, Yingan section; KaFS=Kaipaizileike Formation, Sishichang section; KuFK=Kupukuziman Formation, Keping area; KuFY=Kupukuziman Formation, Yingan section; KuFS=Kupukuziman Formation, Sishichang section; bh: borehole; 1*: major and trace elemental data are from Yu (2009), Hf isotope data are from Li Z.-L. et al. (2012), Sr–Nd isotopic compositions are authors' unpublished data; 2*: Hf isotope data are from Li Z.-L. et al. (2012), the others are from Li Y.-Q. et al. (2012); 3*: Yu (2009) and authors' unpublished Sr–Nd isotopic data; 4*: Hf isotope data are from Li Z.-L. et al. (2012), the others are from Yu et al. (2011), the others are from Yu (2009); 7*: Nd isotope data are from Yu (2009), the others are from Li Z.-L. et al. (2012), Sr isotope ratio is authors' unpublished data, the others are from Yu (2009) ; 9*: Hf isotope data are from Li Z.-L. et al. (2012), the others are from Yu (2009); 10*: major and trace elemental data are from Yu (2009), Hf isotope data are from Li Z.-L. et al. (2012). Nd isotopic compositions are authors' unpublished data; 11*: Hf isotope data are from Li Z.-L. et al. (2012), the others are from Yu (2009)

Supplementary Table I's References

Cao, J., Wang, C.-Y., Xing, C.-M., Xu, Y.-G., 2014. Origin of the early Permian Wajilitag igneous complex and associated Fe–Ti oxide mineralization in the Tarim Large Igneous Province, NW China. Journal of Asian Earth Sciences, 84: 51-68.

Li, H.-Y., Huang, X.-L., Li, W.-X., Cao, J., He, P.-L., Xu, Y.-G., 2013. Age and geochemistry of the Early Permian basalts from Qimugan in the southwestern Tarim Basin. Acta Petrologica Sinica, 29(10): 3353-3368 (in Chinese with English abstract).

Li, Y.-Q., Li, Z.-L., Sun, Y.-L., Santosh, M., Langmuir, C.H., Chen, H.-L., Yang, S.-F., Chen, Z.-X., Yu, X., 2012. Platinum-group elements and geochemical characteristics of the Permian continental flood basalts in the Tarim Basin, Northwest China: Implications for the evolution of the Tarim Large Igneous Province. Chemical Geology, 328: 278-289.

Li, Z.-L., Yang, S.-F., Chen, H.-L., Langmuir, C.H., Yu, X., Lin, X.-B., Li, Y.-Q., 2008. Chronology and geochemistry of Taxinan basalts from the Tarim Basin: Evidence for Permian plume magmatism. Acta Petrologica Sinica, 24(5): 959-970 (in Chinese with English abstract).

Li, Z.-L., Li, Y.-Q., Chen, H.-L., Santosh, M., Yang, S.-F., Xu, Y.-G., Langmuir, C.H., Chen, Z.-X., Yu, X., Zou, S.-Y., 2012. Hf isotopic characteristics of the Tarim Permian Large Igneous Province rocks of NW China: Implication for the magmatic source and evolution. Journal of Asian Earth Sciences, 49: 191-202.

Pan, J.-W., Li, H.-B., Sun, Z.-M., Si, J.-L., Pei, J.-L., Zhang, L.-J., 2011. Geochemistry and possible age of Gudongshan volcanic rocks, Tarim Basin. Geology in China, 38(4): 829-837 (in Chinese with English abstract).

Tian, W., Campbell, I.H., Allen, C.M., Guan, P., Pan, W.-Q., Chen, M.-M., Yu, H.-J., Zhu, W.-P., 2010. The Tarim picrite–basalt–rhyolite suite, a Permian flood basalt from Northwest China with contrasting rhyolites produced by fractional crystallization and anatexis. Contributions to Mineralogy and Petrology, 160(3): 407-425.

Wei, X., Xu, Y.-G., Feng, Y.-X., Zhao, J.-X., 2014. Plume–lithosphere interaction in the generation of the Tarim Large Igneous Province, NW China: Geochronological and geochemical constraints. American Journal of Science, 314: 314-356.

Yu, J.-C., Mo, X.-X., Yu, X.-H., Dong, G.-C., Fu, Q., Xing, F.-C., 2012. Geochemical characteristics and petrogenesis of Permian basaltic rocks in Keping area, western Tarim Basin: A record of plume–lithosphere interaction. Journal of Earth Science, 23(4): 442-454.

Yu, X., 2009. Magma Evolution and Deep Geological Processes of Early Permian Tarim Large Igneous Province. Ph. D. Thesis, Zhejiang University (in Chinese with English abstract).

Yu, X., Chen, H.-L., Yang, S.-F., Li, Z.-L., Wang, Q.-H., Lin, X.-B., Xu, Y., Luo, J.-C., 2009. Geochemical features of Permian basalts in Tarim Basin and compared with Emeishan LIP. Acta Petrologica Sinica, 25(6): 1492-1498 (in Chinese with English abstract).

Yu, X., Yang, S.-F., Chen, H.-L., Chen, Z.-Q., Li, Z.-L., Batt, G.E., Li, Y.-Q., 2011. Permian flood basalts from the Tarim Basin, Northwest China: SHRIMP zircon U–Pb dating and geochemical characteristics. Gondwana Research, 20(2-3): 485-497.

Zhang, C.-L., Li, X.-H., Li, Z.-X., Ye, H.-M., Li, C.-N., 2008. A Permian layered intrusive complex in the Western Tarim Block, northwestern China: Product of a ca. 275-Ma mantle plume. Journal of Geology, 116(3): 269-287.

Zhang, D.-Y., Zhou, T.-F., Yuan, F., Jowitt, S.M., Fan, Y., Liu, S., 2012. Source, evolution and emplacement of Permian Tarim Basalts: Evidence from U–Pb dating, Sr–Nd–Pb–Hf isotope systematics and whole rock geochemistry of basalts from the Keping area, Xinjiang Uygur Autonomous Region, Northwest China. Journal of Asian Earth Sciences, 49: 175-190.

Zhang, Y.-T., Liu, J.-Q., Guo, Z.-F., 2010. Permian basaltic rocks in the Tarim basin, NW China: Implications for plume–lithosphere interaction. Gondwana Research, 18(4): 596-610.

Zhou, M.-F., Zhao, J.-H., Jiang, C.-Y., Gao, J.-F., Wang, W., Yang, S.-H., 2009. OIB-like, heterogeneous mantle sources of Permian basaltic magmatism in the western Tarim Basin, NW China: Implications for a possible Permian Large Igneous Province. Lithos, 113(3-4): 583-594.

4

Geodynamics of the Tarim LIP

Abstract: The Tarim Large Igneous Province (Tarim LIP) agrees well with the mantle plume model in terms of its tempo-spatial distribution of Permian igneous rocks. The domal uplift of the crust before eruption, widespread dike swarms and occurrence of picrate also support the plume involvement in the origin of the Tarim LIP. The inter-relationship between the Tarim LIP and adjacent Permian magmatisms, like the Tianshan Group, the Emeishan LIP and Siberian Traps, indicates that the Tarim LIP is geologically independent from these igneous provinces. The geochemical signatures and tempo-spatial distribution of the Tarim LIP appeals to a two-stage melting model for Permian Tarim magmatism. The geodynamic settings of the Tarim Block in the Early Permian lead to the plume-lithosphere interaction for the petrogenesis of the Tarim LIP.

Keywords: Domal uplift; Dike swarm; Geochemical comparison; Mantle plume; Geodynamic model; Tarim Large Igneous Province (Tarim LIP)

4.1 Relationship Between the Tarim LIP and the Mantle Plume

The Tarim LIP is most readily considered as a result of mantle plume beneath the Tarim Block in the Early Permian (Tian et al., 2010; Yang et al., 2013; Zhou et al., 2009; Pirajno et al., 2008; Zhang CL et al., 2008), just as most of the LIPs around the world which are related to plumes, such as the Karoo Igneous Province (Ellam and Cox, 1991; Jourdan et al., 2007; Hastie, 2014), the Columbia River flood basalt (Hooper, 1997), the North Atlantic Igneous Province (Saunders et al., 1997), the Ontong Java Plateau (Neal et al., 1997), the Parana–Etendeka Province (Peate, 1997), the Siberian Traps (Fedorenko et al., 1996; Sharma, 1997), and Emeishan LIP (Zhou et al., 2002; Xu et al., 2004, 2007). However, other propositions have been suggested as the mechanism for LIP origin, such as plate tectonics (Anderson, 2000) and meteorite impacts (Ingle and Coffin, 2004; Jones, 2005). Coltice et al. (2007) attributed the Early Jurassic (200 Ma) Central Atlantic Magmatic Province (CAMP) to long-term shallow mantle heating associated with continent aggregation. Ingle and Coffin (2004) suggested that the Ontong Java oceanic plateau may be the result of a bolide impact. Ivanov (2007) suggested that the Siberian Trap magmatism is an upper-mantle plate tectonic process related to subduction.

For the Tarim LIP, there are several evidences consistent with the characteristics summarized by Campbell (2001) for identification of ancient mantle plumes. The proposed features of plume-induced magmatism are described as an uplift of the crust before the large-scale volcanism, radiating dike swarms, physical characteristics of volcanism, time-progressive chains of volcanic extrusive bodies (e.g. seamounts or islands), and the chemical composition of the mantle-sourced magma (Campbell, 2001). For the Tarim LIP, crustal uplift just before the large volume magma extrusion has been documented (Li D.-X. et al., 2014; Chen et al., 2006). In addition, the widely developed basic dike swarms and coeval mafic-ultramafic intrusions have been reported (Yang et al., 2007). The occurrence of picrite, though debatable in its nomenclature, is indicative of mantle plume involvement for the genesis of the Tarim LIP (Tian et al., 2010).

4.1.1 Domal Uplift of the Crust Before the Major Eruption

It is apparent that the crust/lithosphere will probably uplift due to the underplating of hot plume (Campbell and Griffiths, 1990; Saunders et al., 2007). This has been proved by many examples of plume-induced LIPs. The crust of the North Atlantic Ocean is supposed to have been uplifted ~54 Ma ago due to the mantle plume upwelling, which might have led to the breakup of Laurasia (Saunders et al., 2005). Smith and Braile (1993) proposed a transient topographic high centered on Yellowstone with a radius of 200 km, suggesting ~1000 m vertical uplift. Saunders et al. (2005) believed that regional uplift occurred in the West Siberian Basin during the Late Permian–Triassic due to the uprising of the Siberian plume head, which caused the rifting of the basin. He et al. (2003) reported the domal uplift immediately preceding the eruption of Emeishan LIP by systematic biostratigraphic correlation and examination of the strata just underneath the basalts.

Detailed stratigraphic work regarding the sediments temporally related to the Tarim LIP has been carried out recently (Li D.-X. et al., 2014; Chen et al., 2006). It reveals that there was a crustal uplift event at the end of the Carboniferous period (predating the Pre-Eruption Deposit Member (PEDM)) for the Tarim Block. The uplift is approximately dome-shaped, with a vertical and lateral extent of more than 887 m and ~300 km, respectively. The center of the dome is in the region of northern Tarim.

● **Carboniferous–Permian Stratigraphy**

Carboniferous strata in the Tarim Basin include 4 formations, i.e. Donghetang, Bachu, Kalashayi and Xiaohaizi (Fig. 4.1), which can be further divided into 6 members (C1–C6) on the basis of their lithology (Zhou et al., 2001; Jia et al., 2004; Guo et al., 2010).

The Permian strata have been assigned different names according to their locations (Jia et al., 2004). For simplicity, we refer to all of these synchronous formations in between the Carboniferous and the Permian igneous rocks as a PEDM (Fig. 4.2). The thickness of the PEDMs generally thins out to the north in the Ma5–Donghe6 profile. This is in accordance with its deposition setting, as deltaic environment in the north and the platform and littoral facies in the south (Fig. 4.2A). The similar deposition patterns can be traced in the Sishichang-Mancan1 profile, i.e. thinning out northwestward (Fig. 4.2B). The distribution of residual PEDMs shows that the thinnest deposition region is located to the east of Keping and west of northern Tarim (Fig. 4.3B).

● **Variable Denudation of the Carboniferous Strata**

Regional stratigraphic correlation has been established to further constrain the spatial distribution of the Carboniferous–Permian strata and examine the denudation process quantitatively, based on marked layers, index fossils and absolute isotopic dating on volcanic rocks (Jia et al., 2004; Zhang et al., 2003). Based on data from the boreholes and outcrops and seismic profiles, the denudation of the Carboniferous strata can be classified into three levels: deep, partial and weak erosion (Fig. 4.3A). The heavier denudation occurred in the

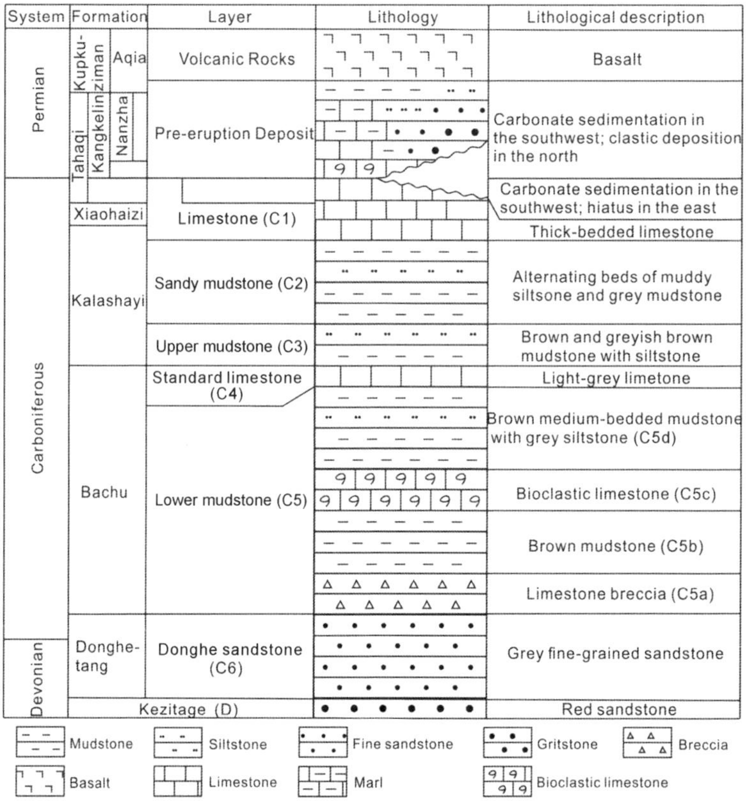

Fig. 4.1 Stratigraphy and lithofacies of the Devonian–Permian sediments in the Tarim Basin (modified from Jia et al., 2004). The gray boxes to the bottom of the Permian strata and to the top of the Carboniferous strata indicate the absent strata in the Zisong Stage and the Xiaoyiaoan Stage, respectively

northwest part of the Basin, which is also evident from the connecting-well stratigraphic profiles (Fig. 4.2).

In addition to the stratigraphic evidence, the noticeable unconformity has also been documented by the absence of significant strata. The Xiaoyiaoan Stage and top Dalan Stage (in the Late Carboniferous), as well as the overlying Nanzha Formation, are absent regionally. This is supported by the absence of fossil zones of *Pseudoschwagerina* (representing the Zisong Stage), *Triticites* (representing the Xiaoyiaoan Stage) and *Protriticites* in drill wells Qun4 and Jia1 in the western central part of the Tarim Basin. The unconformity can be caused by either a simple depositional hiatus or significant erosion. The existence of dissolved pores,

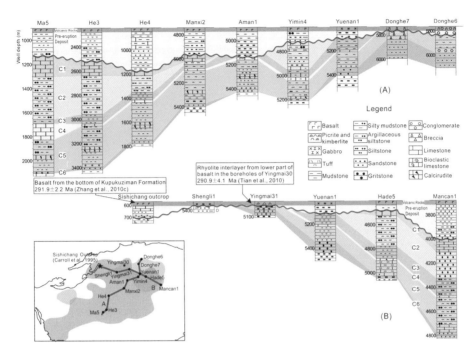

Fig. 4.2 Stratigraphic profiles along the Ma5-Donghe6 (A) and the Sishichang outcrop-Mancan1 profile (B). The locations of the profiles are shown as red lines in the inset map. Data on the Sishichang outcrop was modified after Carroll et al. (1995). The geochronology data of the basalt at the Sishichang outcrop and of rhyolite at Yingmai 30 are after Zhang D.-Y. et al. (2010) and Tian et al. (2010), respectively

caves and other features in boreholes and paleosol layers in the outcrops supports the later interpretation of denudation.

•Late Carboniferous–Early Permian Crustal Uplift

The distribution of denudation and deposition reveals the regional tectonic and stratigraphic evolution during the Late Carboniferous–Early Permian. During the Carboniferous period, mainly carbonate platform sediments were deposited in the Tarim Basin, i.e. C1–C6 (Fig. 4.4A). At the end of the Carboniferous, a crustal doming uplift, occurred rapidly following the deposition of the C1 member. This led to a widespread denudation to the Carboniferous strata, and the absence of an expected biostratigraphic unit containing Fusulinids of Triticites. The C1–C6 members were greatly eroded in the western part of northern Tarim, which is supposed to be the central part of the dome (Fig. 4.4B). After the significant

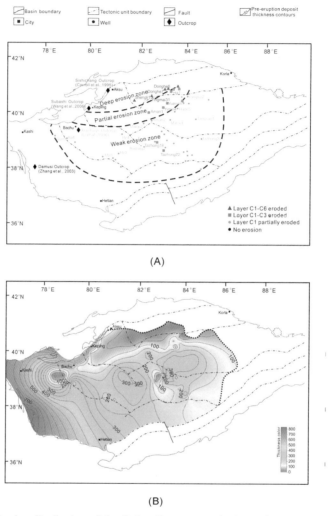

Fig. 4.3 (A) Erosion distribution of the Carboniferous strata in the Tarim Basin; (B) Thickness distribution of the Early Permian Pre-Eruption Deposit Member in the Tarim Basin. No available borehole/outcrop data constrain regions outside the dotted line. The data of four outcrops (Sishichang, Subashi, Xiaohaizi and Damusi) are after Carroll et al. (1995), Wang et al. (2006), Li et al. (1996) and Zhang et al. (2003), respectively

crustal uplift, the regions underwent a slight subsidence, allowing for the deposition of a PEDM with its facies varying from a delta in the north gradually to a platform setting in the southwest. The facies variation of the PEDM suggests the correspondence between the denudation and the deposition (Fig. 4.4C). The most denuded regions, located to the east of Keping and to the west of Northern Tarim, correspond to the thinnest deposition of the PEDM.

The vertical scale of the crust uplift can be estimated by subtracting the smallest thickness of relict Carboniferous strata from the most intact ones (Fig. 4.2). The calculated vertical uplift scale in the Ma5–Donghe6 profile is about 887 m and in the Sishichang–Mancan1 profile about 769 m (Fig. 4.2B). These results should represent a minimum estimation of the scale of the vertical uplift as the uppermost Carboniferous C1 member may not be reserved intact and the effect of depositional compaction is excluded from the calculation. Thus the scale of the crustal uplift could be expected as around 1 km.

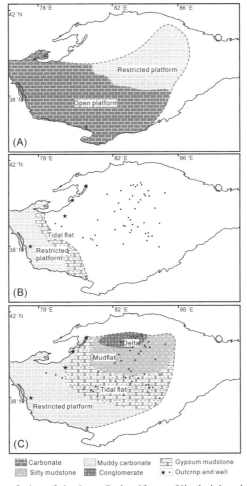

Fig. 4.4 Sedimentary facies of the Late Carboniferous Xiaohaizi and Xiaoyiaoan formations to the Early Permian deposit in the Tarim Basin. (A) Xiaohaizi Formation deposition; (B) Xiaoyiaoan Stage; (C) Early Permian Pre-eruption deposition. (A) is modified from Zhu et al. (2002); (B) and (C) are compiled from borehole and outcrop data. See Fig. 4.3A for the borehole and outcrop names

4.1.2 Widespread Dike Swarms

Giant radiating dike swarms can be attributed to mantle plumes. The Mackenzie swarm in Canada (a giant radiating swarm covering 100° of azimuth), the Central Atlantic reconstructed swarm (a 360° giant radiating swarm), the Matachewan swarm in Canada (deformed giant radiating swarms), the Grenville swarm in Canada and the Yakutsk swarm in Siberia (rift-arm associated swarms) are all typical examples thought to be related to mantle plumes (Ernst et al., 1995; Kiselev et al., 2012).

Outcrops of Permian mafic dike swarms are also widely seen in the Tarim Block. These mafic dike swarms, together with coeval basalts, mafic–ultramafic intrusions and A-type granitic plutons, constitute the Permian Tarim LIP (Yang et al., 2006, 2007; Zhang et al., 2008).

Chen N.-H. et al. (2014) have identified the geometry and emplacement mode of the mafic dikes, sills and flood basalts in Keping and its adjacent areas, including Bachu (northwestern Tarim Block), Beishan and Kuluketag (northeastern Tarim Block) and Kelamyi (western Junggar Block) based on multi-source high-resolution satellite images (Fig. 4.5). Structural analysis and paleomagnetic interpretations were carried out to restore the primary geometry of the mafic dikes. The results indicate that the lengths and thicknesses of 117 extracted mafic dikes in Keping exhibit negative exponent size distributions. The mafic dikes have a mean thickness of 3.8 m with a maximum of 21.4 m. The length of a single mafic dike segment ranges from 0.127 km to 17.1 km. However, the restored geometry of the mafic dikes indicates that the Permian mafic dike swarms in Keping may extend as far as 61–69 km along a primary orientation of about N320°W. The mafic dike swarms and sills in the Tarim LIP make up the plumbing system of the mantle plume. Accompanied by the upwelling of the mantle plume, many eruptive centers and regional dike swarms were generated in the Tarim Basin (Figs. 4.5, 4.6 and 4.7). Dike swarms would indicate a large-scale crustal extension in the Tarim Basin during the Permian age.

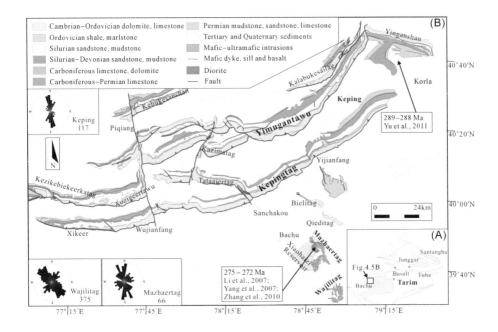

Fig. 4.5 Dike swarms outcropped in NW Tarim Basin (Chen N.-H. et al., 2014)

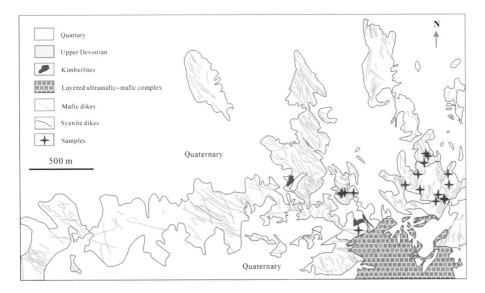

Fig. 4.6 Mafic–ultramafic dike swarms in the Wajilitag area (After Zhang C.-L. et al., 2010)

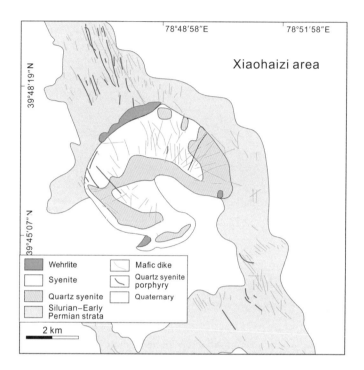

Fig. 4.7 Felsic–mafic–ultramafic dike swarms in the Xiaohaizi region (modified based on Wei et al., 2014b)

4.1.3 Occurrence of Picrite and/or High MgO Igneous Rocks

Mantle plume is thought to be hotter than the ambient asthenosphere. The melts generated from the hotter mantle may show higher MgO concentrations (>12 wt.%). This is supported by the picritic basalt eruption for most of the plume-driven LIPs (Jourdan et al., 2007). Taking the Emeishan LIP as an example, the dominant phenocryst in the picritic flows is Mg-rich olivine (up to 91.6% forsterite component) with high CaO contents (to 0.42 wt.%) and glass inclusions, indicating that the olivine crystallized from a melt. The estimated MgO content of the primitive picritic liquids is about 22 wt.%, and the initial melt temperature might be as high as 1630–1690°C, supporting that the Emeishan LIP can be attributed to the melting of a plume head beginning at approximately 138 km depth (Chung and Jahn, 1995; Zhang et al., 2006). The MgO content of Siberian picrites is up to 18.2 wt.% (Wooden et al., 1993).

The picrite in the Tarim Basin is not so convincing due to its limited surface exposure and insufficient research. The first picrite from the Tarim Basin was officially reported by Tian et al. (2010). Permian picrites were found in the Yingmaili and Yaha areas (YT1, YH5) in the western part of the northern Tarim. Abundant olivine phenocrysts in the picrites provide evidence of crystal accumulation. Alteration of the rocks is manifested by partial to complete replacement of olivine by iddingsite, plagioclase by sericite, glass by chlorite, infilling of vesicles with calcite and ankerite, and by the presence of veins and patches of chlorite and carbonates. The picritic samples have a high amount of MgO (14.5–16.8 wt.%, volatiles included) enriched by an incompatible element and have high $^{87}Sr/^{86}Sr$ and low $^{143}Nd/^{144}Nd$ isotopic ratios ($\varepsilon_{Nd}(t) = -5.3$; $Sr_i = 0.707$), resembling the Karoo high-Ti picrites (Tian et al., 2010).

If the Le Bas (2000)'s definition for picrites (samples with >12 wt.% MgO) is considered, more high MgO rocks in the Tarim Basin can be recognized as picrites. Yang et al. (2007) reported that the geochemical composition of ultrabasic dikes intruded in the Xiaohaizi Formation in the Bachu region is similar to that of picrite, exhibiting contents of MgO to 17% to 22%. These dikes are subvolcanic intrusion, which can be classified as picritic rocks.

4.1.4 Other Features in Favor of the Plume Model

A great many of the metal mineralizations can be associated with plume magmatism, such as copper ore deposits, Cu–Ni sulfide deposits and V–Ti magnetite ore deposits. These ore deposits resulting from mantle plume are usually larger than those formed in a non-plume setting. The well-known Panzihua V–Ti magnetite ore deposit was proved to be associated with the Emeishan LIP and Emeishan plume (Pang et al., 2007). Just as Emeishan, the third largest V–Ti magnetite ore deposit was found in the Wajilitag area in the Tarim Basin. This iron ore resource is estimated to weigh 100×10^6 t, TiO_2 resource 6×10^6 t and V_2O_5 resource 0.13×10^6 t (Gao, 2007). The country rocks for the ore deposit are mainly pyroxenite, gabbro and olivine pyroxenite which are the components in the Tarim LIP (see Chapter 5).

4.2 Geochemical Comparison with other Permian Magmatism in Central Asia

Besides the Tarim LIP, there are two other significant Permian LIPs currently reported in Central Asia, the Emeishan LIP and the Siberian Traps (Fig. 4.8). The Emeishan LIP is now situated in southwestern China, with a distance of about 2300 km from the Tarim LIP in the northwest of China. The Siberian Traps are located in the far north of Asia, 3300 km away from the Tarim LIP, and about 4500 km away from the Emeishan LIP. Recently, Permian magmatism in the Tianshan area (called the Tianshan Group hereinafter) and the north of the Tarim Basin, like the Tuha Basin, the Santanghu Basin and Tianshan Mountain, has been increasingly studied and sometimes related to or even included in the Tarim LIP (Zhang C.-L. et al., 2010). However, before more evidence is verified, we would temporally exclude these Permian magmatisms of the Tianshan Group from the Tarim LIP, as they were probably not geologically settled within the Tarim plate (Yang et al., 2005, 2013). The relationship between these different units of Permian magmatism in Central Asia can be best illustrated in their geochemical compositions. Therefore, the Permian magmatisms of the Tianshan Group, the Emeishan LIP and the Siberian Traps are discussed in terms of their geochemistry for contrasting and relating with the Tarim LIP.

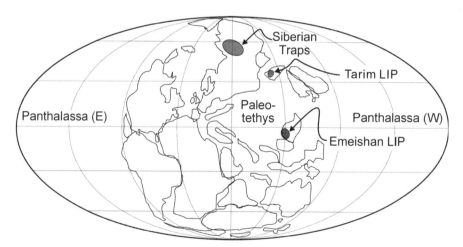

Fig. 4.8 Paleo-latitude of the Tarim LIP, the Emeishan LIP and the Siberian Traps in the Early Permian period (Revised from Ali et al., 2005)

4.2.1 Tarim LIP vs. Tianshan Group Magmatism

For both the Tarim Basin and the Tianshan area, there were a lot of magmatic activities during the Permian period, with various igneous rocks outcropped, such as basalt, andesite, diabase, gabbro, trachyte, rhyolite, volcanic breccia, tuff and so on. For easy comparison, only the basalts were chosen for comparative study in terms of their geochemistry (Fig. 4.9).

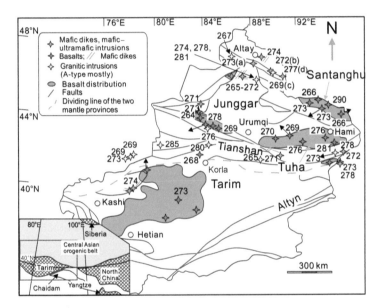

Fig. 4.9 Geotectonic map of the Tarim Block and its adjacent area (Modified after Zhang CL et al., 2010)

- **Major elements**

Generally, the compositions of major elements for Permian basalts from the Tianshan Group (the Tuha Basin, the Santanghu Basin and Tianshan Mountain) are quite similar. The SiO_2 content of the basalts from the Tuha Basin (THB) is around 49.09%–54.40%, TiO_2 about 1.03%–3.49%, the Al_2O_3 about 16.28%–18.38%, the FeO_T about 7.46%–11.54%, the MgO about 4.08%–7.59%, the CaO about 5.64%–10.83%, the Na_2O about 2.61%–4.84%, the K_2O about 0.42%–1.94% and the P_2O_5 about 0.14%–0.54%. The Mg# for the THB is about 0.39–0.59.

The basalts from the Santanghu Basin (STB) have a similar Mg# (0.42–0.59) to that of the THB. All other major elements, such as SiO_2, TiO_2, Al_2O_3,

FeO_T, MgO, CaO, Na_2O and K_2O, are comparable to the THB, except the P_2O_5 (0.35%–1.10 %), which is obviously higher than that for the THB (Zhou et al., 2006).

The basalts from Tianshan Mountain (TSB) have an SiO_2 content and Mg# (0.49–0.52) similar to those of the THB and the STB. However, the concentrations of the major elements are not as scattered as in the THB and the STB according to the existing data. Compared with the THB and the STB, the TSB is relatively lower in TiO_2 (1.50%–1.59%), Al_2O_3 (14.58%–15.11%), Na_2O (2.64%–2.80%), K_2O (0.62%–0.82%) and P_2O_5 (0.20%–0.21 %), but higher in FeO_T (9.84%–10.66%) and CaO (9.09%–9.78%) (Song et al., 2012). Thus, it can be deduced that the THB and the STB are closely related, both of which are in contrast to the TSB in terms of compositions of the major elements (Fig. 4.10).

The basalts reported for the Tarim LIP are with a concentration of SiO_2 of about 40.02–53.6 wt.%, with an average of 45.77 wt.%. MgO=1.96–8.45 wt.%, mean concentration is 4.91 wt.%. TiO_2=1.95–5.25 wt.%, mean concentration is 3.65 wt.%. Al_2O_3=10–17.14 wt.% (mean 13.7 wt.%), CaO=3.35–12.07 wt.% (mean 7.63 wt.%), K_2O=0.35–4.46 wt.% (mean 1.38 wt.%), Na_2O=0.55–5.76 wt.% (mean 3.06 wt.%). The Mg# for Tarim basalts is about 0.22–0.57 (Jiang et al., 2006; Li Z.-L. et al., 2008, 2012; Li Y.-Q., 2012a; Zhang D.-Y. et al., 2012; Yu, 2009; Yu et al., 2011; Zhou et al., 2009).

Compared with Tarim basalts, the Tianshan Group shows a higher content of Al_2O_3, and a lower content of TiO_2, FeO_T and P_2O_5. The Tianshan Group basalts show some intimacy with the Tabei Group from the Tarim LIP, with comparatively higher Mg# than the Keping Group.

- **Trace elements**

From the REE pattern diagram, both the Tianshan Group and Tarim basalts show an enrichment of LREE compared to HREE, but the REE slope for Tarim is much steeper, which means more enrichment of LREE for Tarim basalts than for the Tianshan Group. Besides, the THB has the lowest amount of LREE and the biggest scattering among all the Tarim basalts and the Tianshan Group (Fig. 4.11).

The Tianshan Group is generally rich in LILE as are the Tarim basalts, but with a significant depletion in HFSE (Nb, Ta, Th, U) against Tarim basalts (Fig. 4.12). Santanghu basalts are extremely depleted in Nb and Ta, indicating the involvement of a subduction component in the mantle source (Zhou et al., 2006). Besides, there is a marked positive anomaly in Sr for the Tianshan Group,

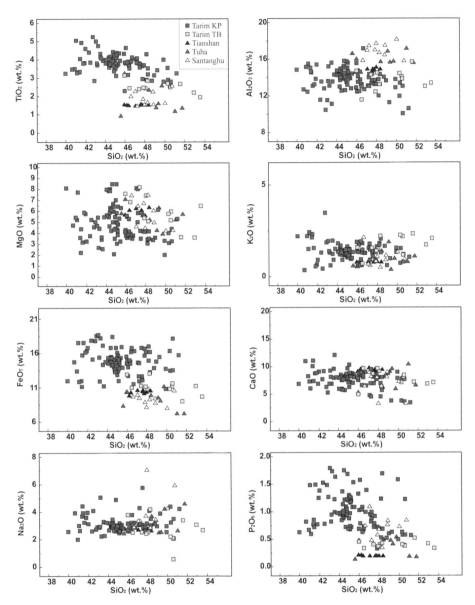

Fig. 4.10 The composition of the major elements for Tarim basalts and the Tianshan Group. The data for the THB and the STB are from Zhou et al. (2006) and Liu et al. (2014). The data for the TSB are selected from Song et al. (2012). The data for Tarim basalts are from Yu (2009), Yu et al. (2011), Jiang et al. (2006), Zhou et al. (2009), Zhang Y.-T. et al. (2010), Zhang D.-Y. et al. (2012) and Li et al. (2008). Only the basaltic samples with MgO <9.0 wt.% were selected within this dataset. The Tarim KP represents the basalts from the Keping Group, while Tarim TB represents Tabei Group basalts. Tianshan reprensents the TSB, Tuha for the THB and Santanghu for the STB

especially those from the Tuha and the Santanghu basins. The study of Middle Permian basalts from the Santanghu Basin shows $Sr/Sr^*=1.44$–2.13, $^{87}Sr/^{86}Sr_i=$ 0.70388–0.70396, $\varepsilon_{Nd}(t)=7.10$–7.99 (Zhao et al., 2006). Among the Tianshan Group, the basalts from Tianshan Mountain show the most intimacy with Tarim basalts, which may indicate the relationship in their petrogenesis.

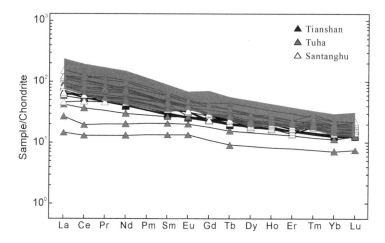

Fig. 4.11 Chondrite normalized REE patterns for Tarim basalts and the Tianshan Group. The dark grey area represents the Keping Group basalts typically outcropped in the Keping area (Yu, 2009). The Keping Group can also be termed as Group 1 as described in Li Y.-Q. et al. (2012a, b)

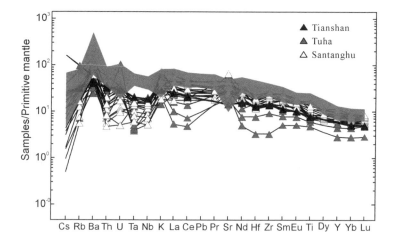

Fig. 4.12 Primitive mantle normalized spider plot for Tarim basalts and the Tianshan Group. The compositions of the primitive mantle are based on MacDonough and Sun (1995). The shaded area represents the trace element composition of the Keping Group basalts from the Tarim LIP

4.2.2 The Tarim LIP vs. the Emeishan LIP

The Tarim LIP and the Emeishan LIP are both located in the western part of China with a time interval of ~30 Ma, and a current distance of ca. 2300 km. Comparing the geochemical features of the basalts from these two LIPs, the Permian basalts from Tarim have geochemical characteristics that are similar to Emeishan basalts, especially the HT Emeishan basalts.

•Major elements

The basalts from the Emeishan LIP can be classified into two groups, LT and HT, with Ti/Y ratios as an indicator (Xu et al., 2001). The HT basalts have a higher amount of TiO_2, FeO_T and P_2O_5, and a lower amount of CaO, Al_2O_3 and MgO, compared with LT basalts (Fig. 4.13).

From the Harker diagrams for the Tarim LIP and the Emeishan LIP basalts, it is obvious that the basalts from the Tarim LIP are mostly equivalent to those of HT from the Emeishan LIP (Fig. 4.13). There is not a large quantity of LT type basalts in the Tarim LIP. The Tarim basalts have a lower content of MgO, thus lower Mg# compared with Emeishan HT and LT basalts. However, the P_2O_5 concentration is relatively higher in Tarim basalts than both HT and LT basalts from the Emeishan LIP. The iron is rich in the Tarim basalt (FeO_T=8.97–18.63 wt.%), with an average of 14.46 wt.%. Besides, Tarim basalts have a lower content of CaO compared with Emeishan basalts.

•Trace elements

From the REE patterns and the spider diagram of trace elements for the Tarim LIP and the Emeishan LIP, it is evident that the Tarim basalts are very similar to those of the HT group from the Emeishan LIP (Figs. 4.14 and 4.15). The total REE abundance of Tarim basalts is much higher than that for the Emeishan LT basalts, even slightly higher than the HT group in Emeishan. That is because the HREE for Tarim basalts is relatively higher than Emeishan HT basalts, and the Emeishan HT basalts is relatively rich in REE compared with the Emeishan LT group. Both the HT and LT basalts are rich in LILE as are the Tarim basalts, with no significant depletion in HFSE, such as Nb, Ta, Hf, Th and U (Fig. 4.15).

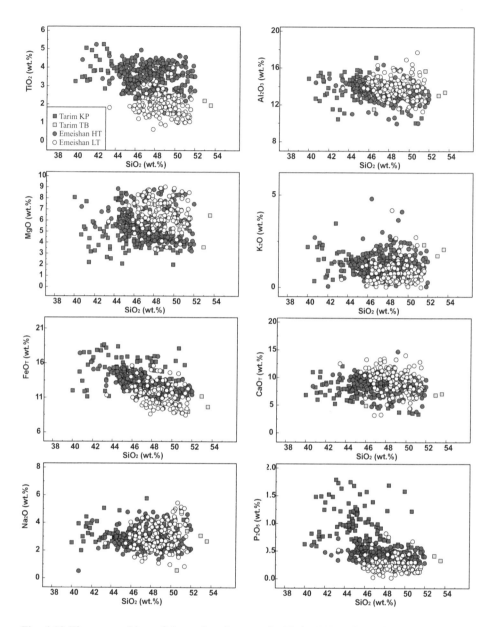

Fig. 4.13 The composition of the major elements for Tarim LIP and Emeishan LIP basalts. Emeishan HT represents high-Ti basalts from the Emeishan LIP, while Emeishan LT represents low-Ti basalts from the Emeishan LIP

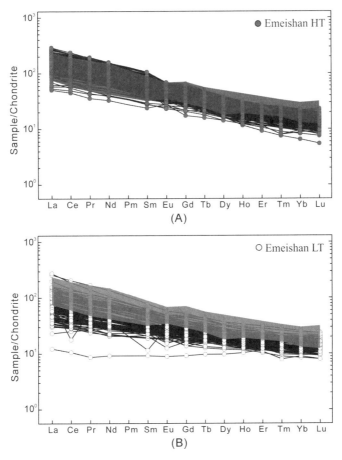

Fig. 4.14 Chondrite normalized REE patterns for (A) Emeishan HT basalts and (B) Emeishan LT basalts, respectively. The shaded area represents the REE composition of the Keping Grop basalts from the Tarim LIP

4.2.3 The Tarim LIP vs. the Siberian Traps

The Siberian Traps are regarded as the largest LIP on land, i.e. up to 3.9×10^6 km^2 (Reichow et al., 2002; Renne, 2002). There are 11 basalt suites which can be grouped into 2 sequences. The lower sequence consists of sub-alkalic basalts, tholeiites and locally picritic basalts of the Ivakinsky (Iv), Syverminsky (Sv) and Gudchichinsky (Gd) formations. The upper sequence includes tufts, tholeiitic and locally picritic basalts of eight formations, i.e., the Hakanchansky (Hk), Tuklonsky (Tk), Nadezhdinsky (Nd), Morongovsky (Mr), Mokulaevsky (Mk), Harayelakhsky (Hr), Kumginsky (Km), and Samoedsky (Sm) formations (Lightfoot, 1993). The

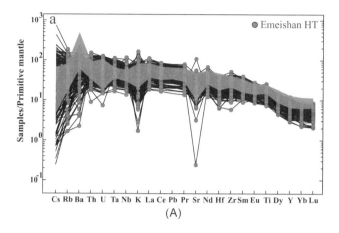
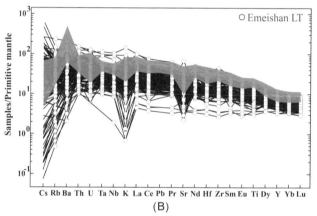

Fig. 4.15 Primitive mantle normalized spider plot for (A) Emeishan HT basalts and (B) Emeishan LT basalts. The shaded area represents the composition of the trace element of Keping Group basalts from the Tarim LIP

estimated thickness of all these suites can add up to 3000 m (Ivanov, 2007).

•Major elements

The basalts can be classified into two geochemical groups, low-Ti and high-Ti, according to their TiO_2-Mg# variation, the Gd/Yb ratio or the Ti/Y ratio (Lightfoot, 1993). The low-Ti basalts have a lower TiO_2 content, and lower Gd/Yb and Ti/Y ratios compared to the high-Ti basalts. Gd/Yb=2.0 can be selected as the boundary that separates low-Ti (<2) and high-Ti (>2) basalts (Lightfoot, 1990).

The lower sequence with three formations (Iv, Sv and Gd) are mainly of the high-Ti type, which has a high TiO_2 content, Ti/Y>410, Gd/Yb>2; while the upper sequence (Hk, Tk, Nd, Mr, Mk, Hr, Km and Sm) is of low-Ti, which makes up to 80% of all the lava sequence of the Siberian Traps. The Tarim LIP, on the contrary, is mainly composed of high-Ti basalts rather than low-Ti.

As the Siberian Traps have so many formations and lava flows, its element concentrations usually extend to a wider range than the Tarim LIP, almost in any of the major elements. So it is more difficult to compare with the Tarim LIP in the overall geochemical features other than in specific sequence or formations. However, it is obvious from the plots that large groups of Siberian Traps are those with lower TiO_2 compared with the Tarim LIP. The SiO_2 content of Siberian Traps basalts is around 45.01–52.00 wt.%, TiO_2 is about 0.39–5.46 wt.%, Al_2O_3 about 8.72–19.16 wt.%, FeO_T about 6.37–16.94 wt.%, MgO about 2.81–18.2 wt.%, CaO about 2.53–17.5 wt.%, Na_2O about 0.15–4.86 wt.% and K_2O about 0.04–5.05 wt.%. The Mg# for the HT type is about 0.30–0.79 and for the LT type is 0.50–0.80. So in comparison to the Tarim LIP, the Siberian Traps, especially the low-Ti type, have lower amounts of TiO_2, FeO_T and P_2O_5, and higher amounts of SiO_2, Al_2O_3 and CaO (Fig. 4.16).

● **Trace elements**

The trace elements and the REE patterns for the Siberian Traps show the wide range of concentration for each element. The HREE ranges of concentration are comparable with those for the Tarim LIP, while the LREE ranges for the Siberian Traps are much wider than those for the Tarim LIP. The LREE for the Tarim LIP is close to the top of the LREE range for the Siberian Traps, which means there are not any counterparts from the Tarim LIP with a low LREE as are most of the low-Ti Siberian basalts (Fig. 4.17).

Generally, the low-Ti basalts are less differentiated between LREE and HREE, with relatively low La/Sm and almost flat REE patterns, which is quite different from those of the Tarim LIP. The low-Ti basalts have a lower abundance of total REE compared with high-Ti and the Tarim LIP. The low-Ti upper sequence basalts are probably from magmas strongly influenced by material from the continental lithosphere, whereas the high-Ti lower sequence basalts are likely generated from a deeper asthenospheric mantle-plume, as those for oceanic island basalts (Lightfoot, 1993). This hypothesis is consistent with the findings on

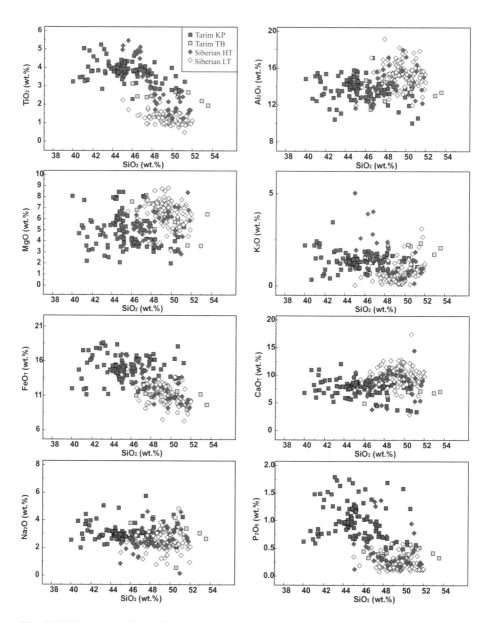

Fig. 4.16 The composition of the major elements for Tarim LIP and Siberian basalts (Siberian data from Hawkesworth et al., 1995; Lightfoot, 1990, 1993; Wooden et al., 1993). Siberian HT represents high-Ti basalts from the Siberian LIP, while Siberian LT represents low-Ti basalts from the Siberian LIP

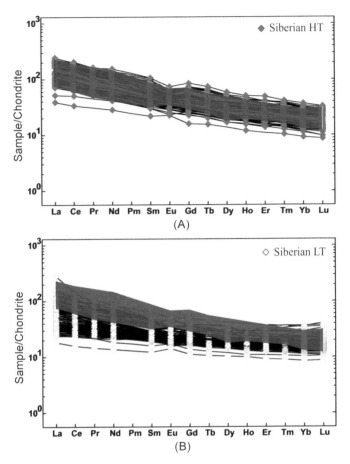

Fig 4.17 Chondrite normalized REE patterns for (A) Siberian HT basalts and (B) Siberian LT basalts. The shaded area represents the REE composition of the Keping Grop basalts from the Tarim LIP

trace elements of the Siberian Traps. The basalts from the upper sequence have a significant negative Ta and Nb anomaly (Nb/La=0.42–0.57), while there is no significant Nb or Ta anomaly (Nb/La=0.8–1.1) for the lower sequence (Fig. 4.18).

From the Ti–Zr diagram (Fig. 4.19), it is apparent that the LT type basalts from the Siberian Traps have a tendency towards island–arc lavas, which does not happen to the Tarim LIP, nor to the Keping Group, or to the Tabei Group. Recent research by Ivanov (2007) suggests that the Siberian Traps magmatism may be related to subduction. The subducting slabs brought significant amounts of water into the mantle transition zone, and consequently releasing water into the upper

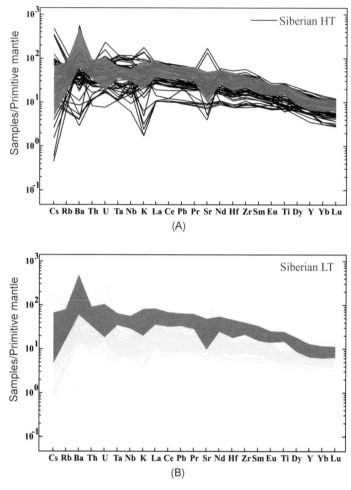

Fig 4.18 Primitive mantle normalized spider plots for (A) Siberian HT basalts and (B) Siberian LT basalts. The shaded area represents the trace element composition of the Keping Group basalts from the Tarim LIP

mantle, lowering the solidus and leading to voluminous melting, which means a lower-mantle plume is not necessary for the origin of the Siberian Traps, but an upper-mantle plate tectonic process will do that. This model may explain what is unexplainable by plume models, such as the inconsistent uplift and subsidence pattern and the absence of mixing curves between expected high-Ti primary plume melts and contaminated low-Ti melts.

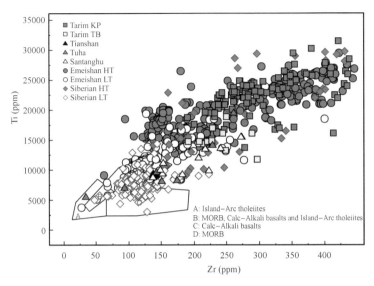

Fig 4.19 The variation of Ti versus Zr for Tarim LIP basalts and other Permian basalts in Central Asia. The discrimination model of tectonic settings is after Pearce and Cann (1973)

4.2.4 Summary of the Permian Magmatism in Central Asia

In Central Asia there are three LIPs currently acknowledged, the Emeishan LIP, the Tarim LIP and the Siberian Traps. In terms of time sequence, the Tarim LIP was extruded in the Early Permian (~290 Ma; Yu et al., 2011; Wei et al., 2014a), the Emeishan in ~260 Ma (Zhou et al., 2002; Guo et al., 2004; He et al., 2007), whereas the Siberian Traps outcropped around 250 Ma, near the Permo–Triassic boundary (Renne and Basu, 1991; Ivanov, 2007). Besides, there are recently some reports about Permian magmatism (293–266 Ma) in the Tianshan area, including the Tuha Basin, the Santanghu Basin, the Junggar Basin and Tianshan Mountain (Zhang CL et al., 2010; Song et al., 2012; Zhou et al., 2006).

The generally acknowledged area for the Tarim LIP is about $2.5 \times 10^5 – 3.0 \times 10^5$ km^2 (Yang et al., 2013; Tian et al., 2010), which is comparable to the Emeishan LIP, 2.5×10^5 km^2 (Ali et al., 2005; Chung and Jahn, 1995; Xu et al., 2001). The Siberian Traps, on the other hand, is much larger than the Tarim LIP and the Emeishan LIP, with an estimated area up to 3.9×10^6 km^2 (Renne, 2002).

While comparing all the basalts from every LIP, it is quite obvious that the contents of SiO$_2$ of Tarim KP Group basalts are the lowest among the 9 groups (Table 4.1). The Tarim KP Group and the TB Group, as well as the Emeishan HT and the Siberian HT have the highest TiO$_2$ contents, thus all classified as high-Ti groups (Fig. 4.20). The FeO$_T$ and P$_2$O$_5$ show some extent of positive correlations (Fig. 4.20). The basalts from Tuha and Santanghu have the highest amount of Al$_2$O$_3$ and Na$_2$O and the lowest of FeO$_T$ (Table 4.1).

Table 4.1 The average compositions of the major elements of typical basalts from Tarim and other regions in Central Asia

Type	SiO$_2$	TiO$_2$	Al$_2$O$_3$	FeO$_T$	MnO	MgO	CaO	Na$_2$O	K$_2$O	P$_2$O$_5$
Tarim KP	45.32	3.79	13.72	14.87	0.21	4.91	7.62	3.11	1.36	0.99
Tarim TB	49.51	2.52	13.75	11.10	0.17	5.90	7.72	2.65	1.71	0.40
Tianshan	47.30	1.54	14.83	10.38	0.16	5.92	9.43	2.71	0.79	0.20
Tuha	48.93	1.84	16.19	9.13	0.22	5.45	7.75	3.24	0.97	0.36
Santanghu	48.40	1.90	16.98	9.23	0.18	5.68	6.93	4.06	1.03	0.71
Emeishan HT	47.48	3.30	12.89	12.59	0.20	6.52	8.93	2.57	1.19	0.41
Emeishan LT	48.74	1.82	13.83	11.15	0.18	6.72	9.69	2.85	0.77	0.23
Siberian HT	48.37	2.96	13.75	12.95	0.20	6.51	8.56	2.46	1.28	0.45
Siberian LT	49.36	1.25	15.33	11.16	0.19	6.93	10.26	2.45	0.65	0.21

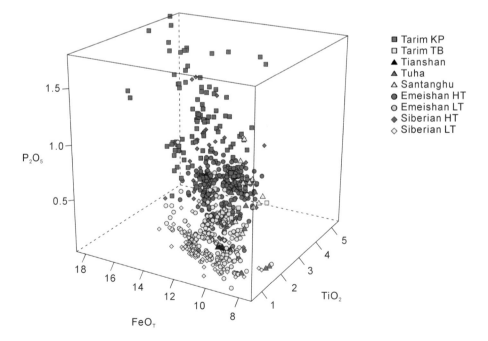

Fig. 4.20 The 3D plot of TiO$_2$–FeO$_T$–P$_2$O$_5$ for Tarim basalts as well as for other Permian basalts in Central Asia

As to the petrogenesis of each LIP, the models are debatable. Ivanov (2007) attributed the Siberian Traps to mostly subduction-related magmatism. The Emeishan LIP may also involve a subduction component in its petrogenesis (Zhu et al., 2005). However, most of the researchers hold the opinion that the Siberian Traps, Emeishan LIP and Tarim LIP are each caused by different mantle plumes (Santosh et al., 2009).

During 300–250 Ma, the three blocks upholding the three LIPs were located around the Paleo–Tethys Ocean far from each other according to the paleomagnetic model. The Tarim Block was located in the latitude of 28°N–30°N (Li et al., 2001), while the Siberian block was further north, and the Emeishan (Yangtze block) was in the south close to the equator (Fig. 4.8). They were all moving northward, but without any cross-over during that period, which means that it is impossible for a single plume to have given birth to all three LIPs.

4.3 Geodynamic Model of the Tarim LIP

The Tarim LIP, just as other Permian LIPs in Central Asia, e.g., the Emeishan LIP and the Siberian Traps, represents a regional tectonic–magmatic event with a very important geodynamic significance, which can best be illustrated by a geodynamic model. Recently, many geologists have paid special attention to the geodynamic model of the Tarim LIP (Yu, 2009; Zhang C.-L. et al., 2010; Wei et al., 2014a; Xu et al., 2014; Yang et al., 2013; Cheng et al., 2015).

Yu (2009) first proposed the two-stage mantle melting model for the Tarim LIP, based on the geochemical difference between the two types of basalts: the Keping Group and the Tabei Group (or Group 1 and Group 2 respectively as referred to in Li Z.-L. et al., 2012). In this model, it was speculated that Group 2 may have been extruded later than Group 1 and the former represents the melting result from plume material, while Group 1 is the result of a partial melting of the lithospheric mantle. However, the existing geochronological data is not supportive for the time sequence of Group 1 and Group 2 basalts, and these two may have been formed during the same period (Tian et al., 2010; Chen et al., 2010).

Zhang C.-L. et al. (2010) suggested that the Tarim basalt is sourced from an enriched lithospheric mantle, while the basic dikes and other intrusions originated from the partial melting of an asthenospheric mantle underneath the lithosphere. This model is consistent with Yu (2009)'s observation; however, there are no details about the petrogeneses of the two types of basalts. The mantle source for these two different groups may be different.

Xu et al. (2014) and Wei et al. (2014a) took into account the latest dating results of kimberitic rocks from the Wajilitag area, and provided the three-stage magmatisms for the Tarim LIP. The kimberlite (~300 Ma) was sourced from the bottom of a metasomatized lithospheric mantle, the flood basalts (~290 Ma) from a plume-lithosphere interaction, while the late mafic and felsic intrusions (~280 Ma) may have originated from an asthenospheric mantle. However, this cannot explain the isotopic intimacy between Stage 1 and Stage 3, i.e., the kimberlite and the intrusions share similar Sr–Nd–Pb isotopic signatures.

Cheng et al. (2015) suggested that the mantle source for kimberlite is from the metasomatism of the subduction material into the sub-lithosphere. The flood basalts resulted from the interaction of plume and lithosphere, which is similar to the viewpoint of Xu et al. (2014). All the late intrusions are from the partial melting of the sub-continental lithospheric mantle, while the latest Wajilitag nephelinite is the result of the melting of the carbonate metasomatized asthenospheric mantle.

These models provide a great amount of information on the possible evolutionary history of the Tarim LIP, and picture the basic geodynamic framework for the Early Permian Tarim magmatism and regional tectonic activity. Nevertheless, there are some missing or imperfect details in both of the models. Here we made the best use of these reported models, and assembled the detailed information on the tempo-spatial distribution of igneous rocks, their geochemical signatures, coupled sedimentary and denudation within the basin, and the radial dike swarms, to propose a new and simple model for the Tarim LIP: The Tarim two-stage melting model.

4.3.1 New Model for the Tarim LIP

The overall outputs of the Tarim LIP can be attributed to two stages of magmatism related to the mantle plume. During the early stage, the magmas were from the partial melting of sub-continental lithosphere mantle (SCLM) which was heated by the extreme hot underplating mantle plume, while during the later stage, the magmas came from the partial melting of the plume itself which was greatly decompressed by upwelling. The detailed geodynamic process of the Tarim LIP can be described as follows (Fig. 4.21).

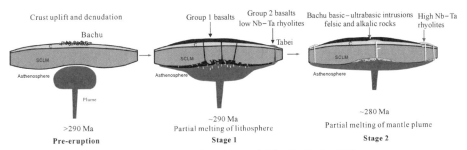

Fig. 4.21 The geodynamic model for the Tarim LIP

(1) Before the Early Permian, the Tarim Block was a cratonic basin which deposited a large sequence of Paleozoic sediments. However, sometime around the Early Permian there had been a mantle plume bred beneath the stable continent lithosphere. The upwelling of the mantle plume caused the uplifting of the overhead continental lithosphere. The lithosphere and crust above the plume head was elevated to a great extent forming the local dome-like structure of the Tarim Block, which is quite probably located in the Bachu area. This would lead to the denudation in these positive landforms.

(2) Meanwhile, the upwelling mantle plume was making contact with the bottom of the continental lithosphere. The continuing impinging of hot plume heated up the SCLM and caused the partial melting of the SCLM. The accumulation of melts enables the extrusion of magma forming the flood basalts widely distributed in the Tarim Block, mainly Group 1 basalts (~290 Ma). On the other hand, the uprising plume migrated horizontally along the bottom of the SCLM. The ongoing migrated plume finally found its melting regime at the periphery of the Tarim Craton, where a thinner lithosphere provided enough room for the plume to decompress adiabatically. At this geological setting and with proper pressure and temperature conditions, the plume would have partially melted to some extent. This melting would have mixed with the melting from the SCLM and contributed to the formation of Group 2 basalts. The fractionation of these magmas might have given rise to the formation of low Nb–Ta type rhyolites (Liu et al., 2014). This is hypothesized to be the Stage 1 magmatism in the Tarim LIP.

(3) With the continued melting of loaded SCLM as well as the ongoing uprising of the plume, the plume head at the center part of the craton also reached the solidus line and began to melt partly. The primitive magma

from the plume would have intruded into the shallow part of the crust with a signature of high magnesium, in the form of ultramafic intrusions. Meanwhile the evolved magma would have given birth to a series of igneous extrusions or intrusions, such as the Wajilitag layered intrusion, diabase, diorite, syenite and nephelinite, etc. At the periphery of the Tarim Basin, such as Tabei, high Nb–Ta rhyolite and felsic intrusions are also outcropped in the boreholes, which were formed by the plume-lithosphere interaction. This is the Stage 2 magmatism in the Tarim LIP.

4.3.2 Evidence Supporting the Model

The geodynamic model is greatly dependent on the geochemical relations among the different components of the Tarim LIP, i.e. ultramafic rocks, basalts of the Keping Group and the Tabei Group, diabases from different parts of the block, syenites and syenitic porphyry, etc.

•Crust uplift and sediment response

A systematic paleo-geographic variation study for the period of the Late Paleozoic suggested that the sedimentary environmental variation is consistent with the LIP development and quite matching to the magmatism center of the Tarim LIP. So the denudation zone of the Tarim Basin in the Early Permian was the reflection of the surface uplift that resulted from the upwelling of the mantle plume (Li et al., 2014; Chen et al., 1996). Li et al. (2014) also indicated the northern part of Tarim as a region with strong denudation, which could be related to the marginal convergence and obduction between the Tarim Block and Tianshan rather than the plume head impingement.

•Tempo-Spatial distribution

The most obvious finding of the temporal relations within the components of the Tarim LIP is that the Keping basalts extruded earlier than the syenites intrusion. The well-acknowledged age for Keping Group basalts is about ~290 Ma (Li Y. et al., 2007; Li Z.-L. et al., 2008; Yang et al., 1996, 2006; Yu, 2009; Yu et al., 2011; Zhang et al., 2003, 2009), while the age for syenites is ~280 Ma (Yang et al., 1996, 2006). The temporal sequence for basaltic rocks and felsic rocks can also be tracked by the strata sequence from boreholes, like Yimin4 with dacite sitting on top of basalt, which means the top one extruded later than the bottom one (Fig. 2.3). According to Li et al. (2011), the sequence of magmatism in the Tarim LIP are basaltic lava in the Keping area, layered mafic–ultramafic rock, mica-olivine pyroxenite breccia pipe, ultramafic dike and diabase, quartz syenite and syenitic porphyry (284–274 Ma), which basically agrees well with the model (Fig. 2.4).

The age of the Tabei Group (Group 2) remains blurry due to the lack of chronological data; however, the rhyolites interlayered with basalts can well limit the time of the Tabei basalts. As reported by Tian et al. (2010), the Tabei Group basalt may have outcropped during the same period as the Keping Group. The distribution of Tabei Group magma is different from that of the Keping Group. The typical one from the Tabei Group is that from boreholes YT6, YM5, YM8 and SL1 in the Tabei area, which are close to the northern periphery of the district of the Keping Group (Fig. 3.1). The distribution of the Tabei Group seems restricted to a certain area other than widely spread. Its location is not perfectly corresponding to the plume head position according to the model, which is consistent with the model.

The most diversified igneous rocks related to the Tarim LIP are outcropped in the Bachu area, including Xiaohaizi and Wajilitag, in the form of mica-olivine pyroxenite breccia pipe, layered mafic–ultramafic rock (pyroxenite, gabbro), ultramafic dike (picritic rocks) and diabase, quartz syenite and syenitic porphyry. The mafic–ultramfic complexes, especially the mica-olivine pyroxenite breccia pipe and the picritic rocks, are consistent with the high-temperature magma that is related to the mantle plume. On the other hand, the Bachu area is the best candidate for the place where the plume head used to be located underneath.

The rhyolites found in the boreholes of the Tabei region can be classified into two stages, with ages that correspond to the two stages of magmatism in the Tarim LIP (Liu et al., 2014).

•Geochemical features

As to the geochemical features for the Tarim LIP, the most significant findings are that there are two types of basalts in the Tarim LIP, called the Keping Group and the Tabei Group, based on their REE patterns. When plotted together in one diagram, it is quite obvious that the basalts from the Tabei Group (Group 2) show a slightly steeper LREE enrichment and more HREE depletion than the Keping Group (Group 1, see Fig. 3.4).

The Keping Group includes basalts from the Keping, Taxinan and

Xiahenan areas and the boreholes He4, TZ22 in the Tazhong area, borehole Ha1 in the northern part of the Tarim Basin. The diabases from the Keping area also display the same characteristics, thus they can be included as a component of the Keping Group. While on the other hand, the Tabei Group mainly includes the basalts from the boreholes YT6, YM5, YM8 and SL1 in the Tabei area. And the diabases from Wajilitag and Xiaohaizi can also be included in the Tabei Group for further discussion in a much simpler way due to their similar geochemical signatures.

The apparent differences in the REE pattern of both of the Group basalts can also be identified from the $(La/Yb)_N$ value, which marks the differentiation between LREE and HREE. The $(La/Yb)_N$ value is 5.5–9.1 for the Keping Group, but 11.1–21.9 for the Tabei Group, which is significantly higher than the former. It is quite certain that those two groups are either from different mantle sources or underwent different magmatic differentiation sourced by the same origin.

The elements Thorium and Ytterbium are of the similar incompatibility during the fractional crystallization of the basaltic magma, so the Th/Yb is considered constant throughout this process which means its ratios can represent the original component of the mantle source. It is evident that the Th/Yb ratios for the two groups are slightly but surely different, with that of the Tabei Group (average 2.54) higher than that of the Keping Group (average 1.29), which indicates that they are from different mantle sources. This is also supported by the Nb/Yb ratios for the two Groups (see Fig. 3.6A). As is shown in Fig. 3.6A, the Nb/Yb ratio for the Keping Group is no less than 15 (adjacent to those for OIB), while the ratio for THE Tabei Group is mostly less than 10 (Li et al., 2014).

Nb/Y values of the two groups are also quite different, which can be used as an indicator of discrimination (Li Y.-Q. et al., 2012a). The ratios of Nb/Y are 0.56–0.82 for the Keping Group (including 58 samples), with an average of 0.68; 1.12–1.74 for the Tabei Group (including 27 samples), with an average of 1.41. The higher Nb/Y values for the Tabei Group indicate that the mantle source for the Tabei Group magma is much more enriched than that for the Keping Group, which is consistent with the involvement of enriched mantle plume for the Tabei Group.

The Sm/Yb–Sm diagram can be used to estimate the degree of partial melting for the mantle source (Fig. 4.22). The high Sm/Yb values (2.2–5.3) indicate that the basaltic magma was from a lower degree of partial melting of the mantle source (Li YQ et al., 2012a). And more precisely, an investigation reveals that there are differences of Sm/Yb values for the two groups, i.e. higher values for the Tabei Group than for the Keping Group. The Tabei Group is therefore from a deeper origin of the mantle (Garnet lherzolite), which is consistent with the model involving a deeper origin of mantle plume for the Tabei Group, whereas it is SCLM (spinel-garnet lherzolite) for the Keping Group. The degree of partial melting would be very low, in the order of 5% (Li YQ et al., 2012a; Yu, 2009; Yu et al., 2011).

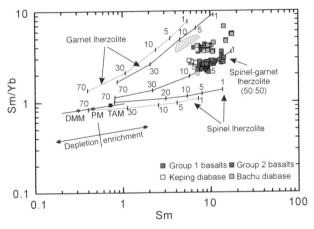

Fig. 4.22 Plot of Sm versus Sm/Yb showing melt curves of the Tarim LIP basalts (Yu et al., 2017). The striated box represents those samples with MgO>7.0 wt.%. The grey area is the composition of picrites (Tian et al., 2010). The melting model is non-model batch melting selected based on Shaw (1970). The partition coefficients for REE are from Mckenzie (1991). PM and N-MORB compositions are from Sun and McDonough (1989). TAM is the best fit mantle source selected for Tarim basalts

Besides, the distinctive Sr and Nd isotopic characteristics for each Group also agree with the model proposed. The Tabei Group (Group 2) has a higher $\varepsilon_{Nd}(t)$ and a lower $^{87}Sr/^{86}Sr_i$ than the Keping Group (Fig. 4.23). The enrichment of the Sr isotope for the Keping Group is attributed to the contamination from the continental crust (Li et al., 2014).

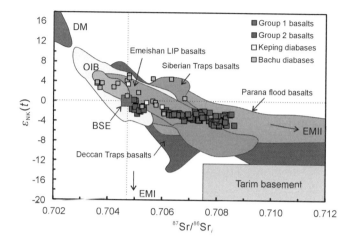

Fig. 4.23 The plot of $^{87}Sr/^{86}Sr_i-\varepsilon_{Nd}(t)$ for the two groups of Tarim basalt as well as diabases. The data for the Tarim basement are from Zhang et al. (2017); The mantle endmembers, like DM, OIB, EMI, EMII, are from Zindler and Hart (1986); The isotopic data for the Emeishan LIP are cited from Chung and Jahn (1995), the Siberian Traps from Wooden et al. (1993), the Deccan Traps from Peng et al. (1998), and the Parana flood basalt from Peate et al. (1990)

4.3.3 Debates on the New Model

Though most of the findings agree well with the geodynamic model suggested in this chapter, there are inevitably more or fewer facts that do not seem exactly supportive of the model or even controversial to it. These facts are also embodied in some specific geochemical features and tempo-spatial distribution for some certain components.

•"Discordant" geochemical features

As the model says, distinct mantle sources are needed for the Keping Group and the Tabei Group respectively. However, the typical indicators like the Zr/Hf ratios and the Nb/Ta ratios, which can indicate the source composition of magma, are almost the same for both of the Groups basalts, with Zr/Hf values of 38.0–54.1 for the Keping Group and 36.2–53.7 for the Tabei Group, Nb/Ta values of 14.7–18.8 and 13.9–18.0 respectively. Though the constant ratios of Zr/Hf and Nb/Ta are not exactly controversial to the model, it is just not supportive either. More research work needs to be carried out to find the factors that may affect the original Zr/Hf or Nb/Ta ratios.

The Nd–Hf isotopic data for the Keping Group and the Tabei Group do not show any marked contrast, which cannot be used to support the model either (Li Z.-L. et al., 2012).

• **"Controversial" tempo-spatial distribution**
The diabases are widely distributed in the northwestern part of the Tarim Basin, such as in the Keping area, the Xiaohaizi area and the Wajilitag area. The existing geochronology data for the diabases are quite ambiguous with a large span of 222–311 million years (Li et al., 2007; Yang et al., 2005). For shortage of robust dating results for diabases, it is difficult to correlate them with certain Groups. What is worse or even more complicated, diabases from different places have distinct geochemical features. The in-depth examination of diabases will help a lot in arguing and verifying the model.

The two types of basalts are distributed in different places with contrast area and volume. The remaining outcrops of these basalts are in Keping, Xiahenan, Taxinan as well as some boreholes in Tazhong, which are all called the "Keping Group". The estimated area of this basaltic extrusion is about 2.5×10^6 km^2, which gives a hint of the diameter of plume head of about 700 km. However, the Tabei Group basalt and the Stage 2 igneous rocks in the Tarim Basin are rather limited. This would be attributed to two factors. The lithosphere for the Tarim Block is very thick, so the decompression melting for plume is limited. On the other hand, the heat conduction to the SCLM from the plume may reduce the internal energy of the plume itself. Thus, the output of Stage 2 magma as well as the Tabei basaltic magma is not well developed.

• **The role of the periphery Permian magmatism for the Tarim LIP**
Pirajno et al. (2009) thought the large-scale magmatism in the Early Permian period also occurred in the eastern part of Kazakhstan and the area between the Siberian craton and the Altai orogen belt. Zhang C.-L. et al. (2010) thought that the Permian mafic and ultramafic magmatism occurred in the Tarim Basin, in the Junggar Basin and in the Turpan–Hami Basin, and that the Altun orogenic belts were parts of the Tarim LIP, with a total area of over 6.0×10^5 km^2 and the main formation time at ca. 275 Ma. Based on the geochemical features of the magmatic rocks, Zhang C.-L. et al. (2010) also subdivided the Tarim LIP into two geochemical provinces: one is the enriched Tarim geochemical province and the

other is the depleted Central Asian Orogenic Belt geochemical province related with the subduction metasomatism. Qin et al. (2011) thought that the Permian basalts of the interior Tarim Basin, Tianshan Mountain, and Beishan Mountain may be derived from the same mantle plume due to the similar geochemical features. However, it should be noted that the popular tectonic evolution model of the Central Asian Orogenic Belt does not support a genetic relationship between the Permian basic and ultra–basic magmatism in northern Xinjiang and the Tarim LIP (Windley et al., 2007). The paleo–Tianshan Ocean that separates northern Xinjiang from the Tarim Block is supposed to have closed westward during the Early Permian.

4.3.4 Summary of the Model

The proposal of this model is inspired by Turner's contribution regarding two types of LIP petrogenesis, i.e. the Parana Pattern for partial melting of the lithospheric mantle and the Deccan Pattern for the partial melting of the mantle plume (Turner et al., 1996). The Tarim LIP is different from both the Parana Pattern and the Deccan Pattern. It is more like the combination of these two patterns. The partial melting of the sub-continental lithospheric mantle comes first, then followed by the melting of the mantle plume itself, leading to the two distinct Groups of basalts in the Tarim LIP, and the further depleted mafic-ultramafic intrusions as well as felsic-alkalic intrusions. The plume-lithosphere interaction is inevitably involved in the petrogenesis of the Tarim LIP, and is just supplementary to the simple Tarim modal.

References

Ali, J.R., Thompson, G.M., Zhou, M., et al., 2005. Emeishan Large Igneous Province, SW China. Lithos, 79: 475-489.

Anderson, D.L., 2000. The thermal state of the upper mantle: No role for mantle plumes. Geophys. Res. Lett., 27(22): 3623-3628.

Campbell, I.H., 2001. Identification of Ancient Mantle Plumes. Mantle plumes: Their identification through times, 352: 5-21.

Campbell, I.H., Griffiths, R.W., 1990. Implications of mantle plume structure for the evolution of flood basalts. Earth Planet Sci Lett, 99: 79-93.

Carroll, A.R., Graham, S.A., Hendrix, M.S., Ying, D., Zhou, D., 1995. Late Paleozoic tectonic amalgamation of northwestern China: Sedimentary record of the northern Tarim, northwestern Turpan, and Southern Junggar Basins. GSA Bulletin, 107(5): 571-594.

Chen, H.-L., Yang, S.-F., Wang, Q.-H., Luo, J.-C., Jia, C.-Z., Wei, G.-Q., Li, Z.-L., He, G.-Y., Hu, A.-P., 2006. Sedimentary response to the Early–Mid Permian basaltic magmatism in the Tarim Plate. Geology in China, 33(3): 545-552 (in Chinese with English abstract).

Chen, M.-M., Tian, W., Zhang, Z.-L., Pan, W.-Q., Song, Y., 2010. Geochronology of the Permian basic–intermediate–acidic magma suite from Tarim, Northwest China and its geological implications. Acta Petrol. Sin., 26(2): 559-572 (in Chinese with English abstract).

Chen, M.-M., Tian, W., Suzuki, K., et al., 2014. Peridotite and pyroxenite xenoliths from Tarim, NW China: Evidences for melt depletion and mantle refertilization in the mantle source region of the Tarim flood basalt. Lithos, 204: 97-111.

Chen, N.-H., Dong, J.-J., Yang, S.-F., Chen, J.-Y., Li, Z.-L., Ni, N.-N., 2014. Restoration of geometry and emplacement mode of the Permian mafic dike swarms in Keping and its adjacent areas of the Tarim Block, NW China. Lithos, 204: 73-82.

Cheng, Z.-G., Zhang, Z.-C., Hou, T., Santosh, M., Zhang, D.-Y., Ke, S., 2015. Petrogenesis of nephelinites from the Tarim Large Igneous Province, NW China: Implications for mantle source characteristics and plume–lithosphere interaction. Lithos, 220-223: 164-178.

Chung, S.-L., Jahn, B.M., 1995. Plume–lithosphere interaction in generation of the Emeishan flood basalts at the Permian–Triassic boundary. Geology, 23: 889-892.

Coltice, N., Bertrand, H., Ricard, Y., Rey, P., 2007. Global warming of the mantle at the origin of flood basalts over supercontinents. Geochimica et Cosmochimica Acta, 70: A108-A108.

Ellam, R.M., Cox, K.G., 1991. An interpretation of Karoo picrite basalts in terms of interaction between asthenospheric magmas and the mantle lithosphere. Earth and Planetary Science Letters, 105: 330-342.

Ernst, R.E., Head, J.W., Parfitt, E., et al., 1995. Giant radiating dike swarms on Earth and Venus. Earth-Sci Rev, 39: 1-58.

Fedorenko, V.I., Lightfoot, P.C., Naldrett, A.J., Czamanske, G.K., Hawkesworth, C.J., Wooden, J.L., Ebel, D.S., 1996. Petrogenesis of the flood-basalt sequence at Noril'sk, North Central Siberia. International Geology Review, 38: 99-135.

Gao, Y.-S., 2007. Geological characteristics of Wajilitag vanadic titanomagnetite deposit and its prospecting recommendations. Xinjiang Iron and Steel, 102: 8-9 (in Chinese).

Guo, F., Fan, W.-M., Wang, Y.-J., et al., 2004, When did the Emeishan mantle plume activity start? Geochronological and geochemical evidence from ultramafic–mafic dikes in southwestern China. Int'l Geol Rev, 46: 226-234.

Guo, J.-H., Sun, S.-D., Xu, J., Guo, C.-Y., Zhang, L.-T., 2010. Sequence stratigraphy of Carboniferous at Bachu area, Tarim Basin. Journal of Central South University: Science and Technology, 41(1): 257-264 (in Chinese with English abstract).

Hastie, W.W., Watkeys, M.K., Aubourg, C., 2014. Magma flow in dike swarms of the Karoo LIP: Implications for the mantle plume hypothesis. Gondwana Res, 25: 736-755.

Hawkesworth, C.J., Lightfoot, P.C., Fedorenko, V.A., Blake, S., Naldrett, A.J., Doherty, W., Gorbachev, N. S., 1995. Magma differentiation and mineralisation in the Siberian continental flood basalts. Lithos, 34: 61-88.

He, B., Xu, Y.-G., Chung, S.-L., Xiao, L., Wang, Y.-M., 2003. Sedimentary evidence for a rapid, kilometer-scale crustal doming prior to the eruption of the Emeishan flood basalts. Earth and Planetary Science Letters, 213: 391-405.

He, B., Xu, Y.-G., Huang, X.-L., et al., 2007. Age and duration of the Emeishan flood volcanism, SW China: Geochemistry and shrimp zircon U–Pb dating of silicic ignimbrites, post-volcanic Xuanwei Formation and clay tuff at the Chaotian section. Earth Planet Sc Lett, 255: 306-323.

Hooper, P.R., 1997, The Columbia river flood basalt province: Current status. In: Mahoney, J.J., Coffin, M.F. (Eds.), Large Igneous Provinces: Continental, Oceanic, and Planetary Flood Volcanism. American Geophysical Union Monograph, 100: 1-27.

Ingle, S., Coffin, M.F. 2004. Impact origin for the greater Ontong Java Plateau. Earth Planet Sc Lett, 218: 123-134.

Ivanov, A.V., 2007. Evaluation of different models for the origin of the Siberian Traps. Geological Society of America Special Papers, 430: 669-691.

Jia, C.-Z., Zhang, S.-B., Wu, S.-Z., 2004. Stratigraphy of the Tarim Basin and Adjacent Areas. Beijing: Science Press, pp. 190-289 (in Chinese).

Jiang, C.-Y., Li, Y.-Z., Zhang, P.-B., Ye, S.-F., 2006. Petrogenesis of Permian basalts on the western margin of the Tarim Basin, China. Russian Geology and Geophysics, 47(2): 232-241.

Jones, A.P., 2005. Meteorite impacts as triggers to large igneous provinces. Elements, 1: 277-281.

Jourdan, F., Bertrand, H., Scharer, U., Blichert-Toft, J., Feraud., G., Kampunzu, A.B., 2007. Major and trace element and Sr, Nd, Hf, and Pb isotope compositions of the Karoo Large Igneous Province, Botswana–Zimbabwe: Lithosphere vs. mantle plume contribution. J Petrol, 48: 1043-1077.

Kiselev, A.I., Ernst, R.E., Yarmolyuk, V.V., Egorov, K.N., 2012. Radiating rifts and dike swarms of the middle Paleozoic Yakutsk plume of eastern Siberian craton. Journal of Asian Earth Sciences, 45: 1-16.

Le Bas, M.J., 2000. IUGS reclassification of the high-Mg and picritic volcanic rocks. J Petrol, 41: 1467-1670.

Li, C.-N., Lu, F.-X., Chen, M.-H., 2001. Research on petrology of the Wajilitag complex body in north edge in the Talimu Basin. Xinjiang Geol., 19: 38-42 (in Chinese with English abstract).

Li, D.-X., Yang, S.-F., Chen, H.-L., et al., 2014. Late Carboniferous crustal uplift of the Tarim Plate and its constraint on the evolution of Early Permian Tarim Large Igneous Province. Lithos, 204: 36-46.

Li, L.-Z., Li, Y.-B., Xiao, C.-T., Liu, B.-L., Jiang, Y.-W., 1996. Carboniferous–Permian

Biostratigraphy in Tarim Basin. Beijing: Geological Press, pp. 1-25 (in Chinese).

Li, Y., Su, W., Kong, P., Qian, Y.-X., Zhang, K.-Y., Zhang, M.-L., Chen, Y., Cai, X.-Y., You, D.-H., 2007. Zircon U–Pb ages of the Early Permian magmatic rocks in the Tazhong–Bachu region, Tarim Basin by LA-ICP-MS. Acta Petrologica Sinica, 23(5): 1097-1107 (in Chinese with English abstract).

Li, Y.-Q., Li, Z.-L., Sun, Y.-L., Santosh, M., Langmuir, C.H., Chen, H.-L., Yang, S.-F., Chen, Z.-X., Yu, X., 2012a. Platinum-group elements and geochemical characteristics of the Permian continental flood basalts in the Tarim Basin, Northwest China: Implications for the evolution of the Tarim Large Igneous Province. Chemical Geology, 328: 278-289.

Li, Y.-Q., Li, Z.-L., Chen, H.-L., Yang, S.-F., Yu, X., 2012b. Mineral characteristics and metallogenesis of the Wajilitag layered mafic–ultramafic intrusion and associated Fe–Ti–V oxide deposit in the Tarim Large Igneous Province, Northwest China. Journal of Asian Earth Sciences, 49: 161-174.

Li, Z.-L., Yang, S.-F., Chen, H.-L., Langmuir, C.H., Yu, X., Lin, X.-B., Li, Y.-Q., 2008. Chronology and geochemistry of Taxinan basalts from the Tarim Basin: Evidence for Permian plume magmatism. Acta Petrologica Sinica, 24(5): 959-970 (in Chinese with English abstract).

Li, Z.-L., Chen, H.-L., Song, B., Li, Y.-Q., Yang, S.-F., Yu, -X., 2011. Temporal evolution of the Permian large igneous province in Tarim Basin in northwestern China. Journal of Asian Earth Sciences, 42: 917-927.

Li, Z.-L., Li, Y.-Q., Chen, H.-L., Santosh, M., Yang, S.-F., Xu, Y.-G., Langmuir, C.H., Chen, Z.-X., Yu, X., Zou, S.-Y., 2012. Hf isotopic characteristics of the Tarim Permian Large Igneous Province rocks of NW China: Implication for the magmatic source and evolution. Journal of Asian Earth Sciences, 49: 191-202.

Lightfoot, P.C., 1990. Geochemistry of the Siberian Trap of the Noril'sk with implications for the relative contributions of crust and mantle to flood basalt magmatism area, USSR. Contrib Mineral Petrol, 104: 631-644.

Lightfoot, P.C., 1993. Remobilisation of the major-, trace-element, and from picritic and tholeiitic Siberian trap, Russia continental lithosphere by a mantle plume: Sr-, Nd-, And Pb-isotope evidence iavas of the Noril'sk district. Contrib Mineral Petrol, 114: 171-188.

Liu, H.-Q., Xu, Y.-G., Tian, W., Zhong, Y.-T., Mundil, R., Li, X.-H., Yang, Y.-H., Luo, Z.-Y., Shangguan, S.-M., 2014. Origin of two types of rhyolites in the Tarim Large Igneous Province: Consequences of incubation and melting of a mantle plume. Lithos, 204: 59-72.

McDonough, W. F., Sun, S., 1995. The composition of the Earth. Chemical Geology, 120(3-4): 223-253.

Neal, C.R., Mahoney, J.J., Kroenke, L.W., Duncan, R.A., Petterson, M.G., 1997. The Ontong Java Plateau. In: Mahoney, J.J., Coffin, M.F. (Eds.), Large Igneous Provinces: Continental, Oceanic, and Planetary Flood Volcanism. American Geophysical Union Monograph, 100: 183-216.

Pang, K.-N., Zhou, M.-F., Lindsley, D., Zhao, D., Malpas, J., 2008. Origin of Fe–Ti oxide ores in mafic intrusions: Evidence from the Panzhihua intrusion, SW China. Journal of Petrology, 49(2): 295-313.

Pearce, J.A., Cann, J.R., 1973, Tectonic setting of basic volcanic rocks determined using trace element analyses. Earth Planet. Sci. Lett., 19(2): 290-300.

Peate, D.W., 1997. The Parana–Etendeka province. In: Mahoney, J.J., Coffin, M.F. (Eds.), Large Igneous Provinces: Continental, Oceanic, and Planetary Flood Volcanism. American Geophysical Union Monograph, 100: 217-246.

Peate, D.W., Hawkesworth, C.J., Mantovani, M.S.M., et al., 1990. Mantle plumes and flood-basalt stratigraphy in the Parana, South America. Geology, 18: 1223-1226.

Peng, Z.-X., Mahoney, J.J., Hooper, P.R., et al., 1998. Basalts of the northeastern Deccan Traps, India: Isotopic and elemental geochemistry and relation to southwestern Deccan stratigraphy. J. Geophys. Res., 103: 29843-29865.

Pirajno, F., Mao, J.-W., Zhang, Z.-C., Zhang, Z.-H., Chai, F.-M., 2008. The association of mafic–ultramafic intrusions and A-type magmatism in the Tianshan and Altai orogens, NW China: Implications for geodynamic evolution and potential for the discovery of new ore deposits. Journal of Asian Earth Sciences, 32: 165-183.

Pirajno, F., Ernst, R.E., Borisenko, A.S., Fedoseev, G., Naumov, E.A., 2009. Intraplate magmatism in Central Asia and China and associated metallogeny. Ore Geology Reviews, 35(2): 114-136.

Qin, K.-Z., Su, B.-X., Sakyi, P.A., Tang, D.-M., Li, X.-H., Sun, H., Xiao, Q.-H., Liu, P.-P., 2011. SIMS zircon U–Pb geochronology and Sr–Nd isotopes of Ni–Cu-bearing mafic–ultramafic intrusions in Eastern Tianshan and Beishan in correlation with flood basalts in Tarim Basin (NW China): Constraints on a ca. 280 Ma mantle plume. American Journal of Science, 311(3): 237-260.

Reichow, M.K., Saunders, A.D, White, R.V., et al., 2002. $^{40}Ar/^{39}Ar$ dates from the West Siberian Basin: Siberian flood basalt province doubled. Science, 296: 1846-1849.

Renne, P.R., 2002. Flood basalts bigger and badder. Science, 297: 522-522.

Renne, P.R., Basu, A.R. 1991. Rapid eruption of the Siberian traps flood basalts at the Permo–Triassic boundary. Science, 253: 176-179.

Santosh, M., Maruyama, S., Yamamoto, S., 2009. The making and breaking of super-continents: some speculations based on superplumes, super downwelling and the role of the tectosphere. Gondwana Res., 15: 324-341.

Saunders, A.D., Fitton, J.G., Kerr, A.C., Norry, M.J., Kent, R.W., 1997.The North Atlantic Igneous Province. In: Mahoney, J.J., Coffin, M.F. (Eds.), Large Igneous Provinces: Continental, Oceanic, and Planetary Flood Volcanism. American Geophysical Union Monograph, 100: 45-94.

Saunders, A.D., England, R.W., Reichow, M.K., White, R.V., 2005. A mantle plume origin for the Siberian Traps: Uplift and extension in the West Siberian Basin, Russia. Lithos, 79: 407-424.

Saunders, A.D., Jones, S.M., Morgan, L.A., Pierce, K.L., Widdowson, M., Xu, Y.-G., 2007.

Regional uplift associated with continental large igneous provinces: The roles of mantle plumes and the lithosphere. Chem. Geol., 241: 282-318.

Sharma, M., 1997. Siberian Traps. In: Mahoney, J.J., Coffin, M.F. (Eds.), Large Igneous Provinces: Continental, Oceanic, and Planetary Flood Volcanism. American Geophysical Union Monograph, 100: 247-272.

Shaw, D.M., 1970. Trace element fractionation during anatexis. Geochimica et Cosmochimica Acta, 34(2): 237-243.

Smith, R.B., Braile, L.W., 1993. Topographic signature, space-time evolution, and physical properties of the Yellowstone–Snake River Plain volcanic system: The Yellowstone hotspot. Geology of Wyoming: Geological Survey of Wyoming Memoir, 5: 694-754.

Song, H.-X., Luo, J.-H., Tang, J.-Y., Zhang, C., 2012. Geochemistry, and its tectonic significances of basaltic magma of Xiaotikanlike Formation in southern margin of south Tianshan, Baichen County. Chinese Journal of Geology, 47(3): 886-898 (in Chinese with English abstract).

Sun, S.-S., McDonough, W.F., 1989. Chemical and isotopic systematics of oceanic basalt. In: Saunders, A.D., Norry, M.J. (Eds.), Magmatism in the Ocean Basins. Geological Society Special Publication, 42: 313-345.

Tian, W., Campbell, I.H., Allen, C.M., et al., 2010. The Tarim picrite–basalt–rhyolite suite, a permian flood basalt from Northwest China with contrasting rhyolites produced by fractional crystallization and anatexis. Contrib Mineral Petr, 160: 407-425.

Turner, S., Hawkesworth, C., Gallagher, K., et al., 1996. Mantle plumes, flood basalts, and thermal models for melt generation beneath continents: Assessment of a conductive heating model and application to the Parana. J Geophys Res, 101: 11503-11518.

Wang, L.-D., Yu, B.-S., Zhang, Y.-W., Miao, J.-J., 2006. Characteristics of reef and beach facies in the Kangkelin age from Western Tarim Basin—A case study from the Subashi outcrop section in the Keping Area. Geoscience, 20(2): 291-298 (in Chinese with English sabstract).

Wei, X., Xu, Y.-G., Feng, Y.-X., Zhao, J.-X., 2014a. Plume–lithosphere interaction in the generation of the Tarim Large Igneous Province, NW China: Geochronological and geochemical constraints. American Journal of Science, 314(1): 314-356.

Wei, X., Xu, Y.-G., Zhang, C.-L., et al., 2014b. Petrology and Sr–Nd isotopic disequilibrium of the Xiaohaizi intrusion, NW China: Genesis of layered intrusions in the Tarim Large Igneous Province. J Petrol, 55: 2567-2598.

Windley, B.F., Alexeiev, D., Xiao, W., Kröner, A., Badarch, G., 2007. Tectonic models for accretion of the Central Asian Orogenic Belt. Journal of the Geological Society, 164(12): 31-47.

Wooden, J.L., Czamanske, G.K., Fedorenko, V.A., et al., 1993. Isotopic and trace-element constraints on mantle and crustal contributions to Siberian continental flood basalts, Noril'sk area, Siberia. Geochimica et Cosmochimica Acta, 57(15): 3677-3704.

Xu, Y.-G., Chung, S.-L., Jahn, B.M., Wu, G.-Y., 2001. Petrologic and geochemical constraints on the petrogenesis of Permian–Triassic Emeishan flood basalts in southwestern China. Lithos, 58: 145-168.

Xu, Y.-G., He, B., Chung, S.-L., Menzies, M.A., Frey, F.A., 2004. Geologic, geochemical, and geophysical consequences of plume involvement in the Emeishan flood-basalt province. Geology, 32: 917-920.

Xu, Y.-G., He, B., Huang, X., Luo, Z., Chung, S.-L., Xiao, L., Zhu, D., Shao, H., Fan, W., Xu, J., Wang, Y., 2007. Identification of mantle plumes in the Emeishan Large Igneous Province. Episodes, 30: 32-42.

Xu, Y.-G., Wei, X., Luo, Z.-Y., Liu, H.-Q., Cao, J., 2014. The Early Permian Tarim Large Igneous Province: Main characteristics and a plume incubation model. Lithos. http://dx.doi.org/10.1016/j.lithos.2014.02.015.

Yang, S.-F., Chen, H.-L., Dong, C.-W., Jia, C.-Z., Wang, Z.-G., 1996. The discovery of Permian syenite inside Tarim Basin and its geodynamic significance. Geochimica, 25(2): 121-128 (in Chinese with English abstract).

Yang, S.-F., Chen, H.-L., Ji, D.-W., Li, Z.-L., Dong, C.-W., Jia, C.-Z., Wei, G.-Q., 2005. Geological process of early to middle Permian magmatism in Tarim Basin and its geodynamic significance. Geological Journal of China Universities, 11(4): 504-511 (in Chinese with English abstract).

Yang, S.-F., Li, Z.-L., Chen, H.-L., et al., 2006. ^{40}Ar-^{39}Ar dating of basalts from Tarim Basin, NW China and its implication to a Permian thermal tectonic event. Journal of Zhejiang University - Science A, 7: 320-324.

Yang, S.-F., Yu, X., Chen, H.-L., Li, Z.-L., Wang, Q.-H., Luo, J.-C., 2007. Geochemical characteristics and petrogenesis of Permian Xiaohaizi ultrabasic dike in Bachu area, Tarim Basin. Acta Petrologica Sinica, 23(5): 1087-1096 (in Chinese with English abstract).

Yang, S.-F., Chen, H.-L., Li, Z.-L., et al., 2013. Early Permian Tarim Large Igneous Province in Northwest China. Science China, Series D: Earth Sciences, 56: 2015-2026.

Yu, X., 2009. Magma Evolution and Deep Geological Processes of Early Permian Tarim Large Igneous Province. Ph. D. Thesis, Zhejiang University (in Chinese with English abstract).

Yu, X., Yang, S.-F., Chen, H.-L., Chen, Z.-Q., Li, Z.-L., Batt, G.E., Li, Y.-Q., 2011. Permian flood basalts from the Tarim Basin, Northwest China: SHRIMP zircon U–Pb dating and geochemical characteristics. Gondwana Research, 20(2-3): 485-497.

Yu, X., Yang, S.-F., Chen, H.-L., Li, Z.-L., Li, Y.-Q. 2017. Petrogenetic model of the Permian Tarim Large Igneous Province. Science China Earth Sciences, 60: 1805-1816.

Zhang, C.-L., Li, X., Li, Z., Ye, H., Li, C., 2008. A Permian layered intrusive complex in the western Tarim Block, northwestern China: Product of a ca. 275-Ma mantle plume. Journal of Geology, 116: 269-287.

Zhang, C.-L., Xu, Y.-G., Li, Z.-X., Wang, H.-Y., Ye, H.-M., 2010. Diverse Permian magmatism in the Tarim Block, NW China: Genetically linked to the Permian Tarim mantle plume. Lithos, 119(3-4): 537-552.

Zhang, D.-Y., Zhou, T.-F., Yuan, F., Fan, Y., Liu, S., Du, H.-X., 2010. LA-ICPMS U–Pb ages,

Hf isotope characteristics of zircons from basalts in the Kupukuziman Formation, Keping area, Tarim Basin. Acta Petrologica Sinica, 26(3): 963-974 (in Chinese with English abstract).

Zhang, D.-Y., Zhou, T.-F., Yuan, F., Jowitt, S.M., Fan, Y., Liu, S., 2012. Source, evolution and emplacement of Permian Tarim basalts: Evidence from U–Pb dating, Sr–Nd–Pb–Hf isotope systematics and whole rock geochemistry of basalts from the Keping area, Xinjiang Uygur Autonomous Region, Northwest China. Journal of Asian Earth Sciences, 49: 175-190.

Zhang, H.-A., Li, Y.-J., Wu, G.-Y., et al., 2009. Isotopic geochronology of Permian igneous rocks in the Tarim Basin. Chin J Geol, 44: 137-158 (in Chinese).

Zhang, S.-B., Ni, Y.-N., Gong, F.-H., Lu, H.-N., Huang, Z.-B., Lin, H.-L., 2003. A Guide to the Stratigraphic Investigation on the Periphery of the Tarim Basin. Beijing: Petroleum Industry Press, pp. 191-217.

Zhang, Y., Wei, X., Xu, Y.-G., et al., 2017. Sr–Nd–Pb isotopic compositions of the lower crust beneath northern Tarim: Insights from igneous rocks in the Kuluketage area, NW China. Miner Petrol, 111(2): 237-252.

Zhang, Y.-T., Liu, J.-Q., Guo, Z.-F., 2010. Permian basaltic rocks in the Tarim Basin, NW China: Implications for plume–lithosphere interaction. Gondwana Research, 18(4): 596-610.

Zhang, Z.-C., Mahoney, J.J., Mao, J.-W., Wang, F.-S., 2006. Geochemistry of picritic and associated basalt flows of the western Emeishan flood basalt province, China. Journal of Petrology, 47: 1997-2019.

Zhao, Z.-H., Guo, Z.-J., Han, B.-F., Wang, Y., 2006. The geochemical characteristics and tectonic–magmatic implications of the latest-Paleozoic volcanic rocks from Santanghu Basin, eastern Xinjiang, Northwest China. Acta Petrologica Sinica, 22(1): 199-214.

Zhou, Z.-Y., Zhao, Z.-X., Hu, Z.-X., Chen, P.-J., Zhang, S.-B., Yong, T.-S., 2001. Stratigraphy of the Tarim Basin. Beijing: Science Press, pp. 119-206 (in Chinese).

Zhou, D.-W., Liu, Y.-Q., Xing, X.-J., et al., 2006. Paleo-tectonic setting restoration and regional structure tracing of Permian basalt in Tuha and Santanghu basins, Xinjiang. Science in China, Series D: Earth Science, 36(2): 143-153.

Zhou, M.-F., Malpas, J., Song, X.-Y., Robinson, P.T., et al., 2002. A temporal link between the Emeishan Large Igneous Province (SW China) and the end-Guadalupian mass extinction. Earth and Planetary Science Letters, 196(3-4): 113-122.

Zhou, M.-F., Zhao, J., Jiang, C., et al., 2009. OIB-like, heterogeneous mantle sources of Permian basaltic magmatism in the western Tarim Basin, NW China: Implications for a possible Permian Large Igneous Province. Lithos, 113: 583-594.

Zhu, B.-Q., Hu, Y. -G., Chang, X. -Y., Xie, J., Zhang, Z.-W., 2005. The Emeishan Large Igneous Province originated from magmatism of a primitive mantle plus subducted slab. Russian Geology and Geophysics, 46: 904-921.

Zhu, R.-K., Luo, P., Luo, Z., 2002. Lithofacies palaeogeography of the late Devonian and

Carboniferous in Tarim Basin. Journal of Palaeogeography, 4(1): 13-24 (in Chinese with English abstract).

Zindler, A., Hart, S.R., 1986. Chemical geodynamics. Annu. Rev. Earth Planet. Sci., 14: 493-571.

5

Metallogenesis of the Tarim LIP

Abstract: The Wajilitag Fe–Ti–V oxide deposit is the only ore body that is being mined in the Tarim LIP. It was probably formed by a slowly cooling magmatic crystallization from a deep magma chamber. Fe–Ti oxides therein were formed by two stages of fractionations: minor Fe–Ti oxides as oxide inclusions in the silicate minerals were crystallized in the early stage of the magmatic process, whereas most of the Fe–Ti oxides were accumulated interstitial to the silicate minerals in the late stage of the magmatic process when silicate crystals had already cumulated. The parental magma of the Wajilitag intrusion was enriched in Fe and Ti, which may be produced by a low degree of partial melting from the mantle source. Extensively fractional crystallization may have caused the evolved magma with a further Fe–Ti enrichment.

The PGE (Platinum-group elements) geochemistry study reveals that the Tarim CFBs are extremely depleted in PGE concentrations, which is mainly due to the very low degrees ($<5\%$) of partial melting in the mantle source. The degree of partial melting plays an important role in the Cu–Ni–PGE mineralization in the Tarim LIP, thus mafic–ultramafic rocks formed by high-degree partial melting from the mantle source in the Tarim LIP should receive more attention in terms of their Cu–Ni–PGE mineralization potential. Besides, the PGE study on the Keping basalts indicates that magma replenishment in the magma chamber may once again trigger S-saturation and saturated sulfide segregation in the evolved magma, which may make it possible to develop a Cu–Ni–PGE sulfide deposit in the Keping area.

Keywords: Wajilitag; Fe–Ti–V oxide deposit; PGE; Metallogenesis

Fe–Ti–V oxides and Cu–Ni–PGE sulfides are important mineral resources, and both of them are often genetically related to the LIPs around the world (Bryan and Ernst, 2008; Naldrett, 2004; Pirajno, 2013). For example, both Fe–Ti oxide deposits and Cu–Ni–PGE deposits have been mined in the Emeishan LIP in SW China (Song et al., 2008; Zhou et al., 2005, 2008). In the Tarim LIP, at least one Fe–Ti–V oxide deposit (the Wajilitag deposit) has been found and exploited (Gao, 2007), whereas the potential of Cu–Ni–PGE deposits in this LIP is still unclear. This chapter will introduce the metallogenesis of the Wajilitag Fe–Ti–V oxide deposit and investigate the Cu–Ni–PGE mineral resource potential in the Tarim LIP.

5.1 Wajilitag Fe–Ti–V Oxide Deposit

The Wajilitag giant Fe–Ti–V oxide deposit is situated in a layered mafic–ultramafic intrusive complex (the Wajilitag complex) in the southeastern part of Bachu County, western Tarim Basin (Fig. 5.1). It first caught the attention of geologists due to its strong negative magnetic anomaly through an aeromagnetic investigation in the 1950s. A further geological and geophysical survey by the Second Geological Party of Xinjiang Bureau of Geology and Mineral Resources found that this intrusion owns considerable Fe–Ti oxides (magnetite and ilmenite) as well as some vanadium mineralization, comparable to the Panzhihua and Hongge mafic–ultramafic intrusions and ore deposits in the Emeishan LIP of SW China (Zhong et al., 2002; Zhou et al., 2005). The intrusion possesses 100 million tons (Mt) of Fe–Ti–V oxide ore reserves with ca. 20 wt.% FeO_T (as total iron), 7 wt.% TiO_2 and 0.14 wt.% V_2O_5 (Gao, 2007), which has the potential of becoming a giant Fe–Ti–V oxide deposit.

The Wajilitag complex, as an important lithological unit of the Tarim LIP, is stock-like with an exposed area of ca. 15 km^2 (Fig. 5.2). It intruded into the Silurian–Devonian sedimentary rocks, with contact zones dipping 20°–40° towards the interior of the complex, and its southern part is covered by a desert. The intrusive complex consists of olivine pyroxenite, pyroxenite and gabbro bodies from the lower part to the upper part and shows rhythmic layered structures. Besides, there are some nepheline-bearing syenites outcropped on the top of the intrusion, and widespread mafic dikes cut through the intrusion and surrounding sedimentary strata.

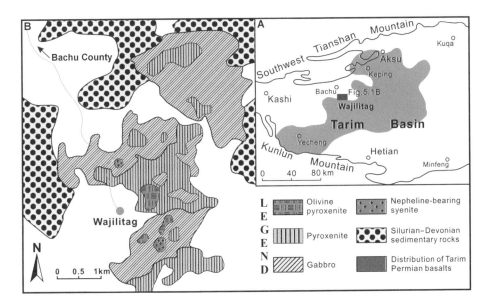

Fig. 5.1 (A) A regional geological map showing the location of the Wajilitag Fe–Ti–V oxide deposit in the Tarim LIP (After Yang et al., 2005); (B) A simplified geological map of the Wajilitag layered mafic–ultramafic intrusive complex (Modified after Zhang et al., 2008)

Fig. 5.2 A panoramic photograph of the Wajilitag intrusive complex. It (outlined by the orange dashed line) intrudes the Silurian–Devonian sedimentary rocks, and the southern part of the intrusion is covered by the Taklamakan Desert. Mafic dikes cut widely through the intrusion and some nepheline-bearing syenites are outcropped on the top

A whole-rock Sm–Nd isochron age of (306±7) Ma was reported for the Wajilitag complex (Lu et al., 2000). However, Zhang et al. (2008) considered that the emplacement age of the intrusion might be ca. 275 Ma based on the zircon U–Pb dating of the syenite in the Bachu area. In addition, some olivine-rich ultramafic dikes with a trend in a north and northwest direction occurred in the Xiaohaizi area (known also as the Mazhaertage or Mazhartag area), north of the Wajilitag intrusion (Jiang et al., 2004; Yang et al., 2007; Zhou et al., 2009).

According to the available geochemical and geochronological data, Zhang et al. (2010) suggested that this mafic–ultramafic intrusion and dikes, along with other igneous rocks in the Wajilitag region, shared the same mantle source and magmatic process, which were probably a product of a Permian upwelling mantle plume under the Tarim Block.

5.1.1 Petrography of the Main Rock Units in the Wajilitag Deposit

The rock units in the Wajilitag deposit mainly include olivine pyroxenite, coarse-grained pyroxenite, fine-grained pyroxenite and gabbro. Their general petrography is as follows:

(1) The olivine pyroxenite, mostly located in the lower part of the Wajilitag complex, is mainly composed of olivine (20–30 modal%), clinopyroxene (50–70 modal%) and Fe–Ti oxides (5–10 modal%) with minor plagioclase and apatite. The grain size of olivine in the olivine pyroxenite is about 3–5 mm, and the grain size of clinopyroxene is about 5–7 mm.

(2) The pyroxenite is predominant in the deposit and contains a large amount of Fe–Ti oxide ores. It can be subdivided into coarse- and fine-grained pyroxenites based on the grain size of clinopyroxenes. The coarse-grained pyroxenite has inequigranular clinopyroxene (ca. 75 modal% in content) ranging in size from 0.5 mm×1 mm to 5 mm×10 mm along with minor olivine, plagioclase and Fe–Ti oxides. The fine-grained pyroxenite contains equigranular clinopyroxene (mostly 0.5 mm×0.5 mm in size, 70–90 modal% in content) filled with interstitial Fe–Ti oxides, forming a net-texture (Fig. 5.3A).

(3) Fine- to middle-grained gabbro sits on the upper part of the Wajilitag complex. It contains ca. 50 modal % clinopyroxene, ca. 40 modal% plagioclase and minor Fe–Ti oxides and hornblende (Fig. 5.3B). Some parts in the gabbro contain apatite up to 5 modal % (Fig. 5.3C).

On the whole, olivine is generally homogeneous in composition and partly replaced by iddingsite in the olivine pyroxenite (Fig. 5.3D). Clinopyroxene in pyroxenites is commonly rimmed with hornblende (Fig. 5.3E). A number of clinopyroxene minerals in gabbros have a compositional zonation from the core to the rim (Fig. 5.3F). The opaque minerals are either interstitial to the silicates or enclosed in the clinopyroxene and/or plagioclase as euhedral to subhedral grains, and a few of them are enclosed in olivines (Fig. 5.3G). They are mostly Fe–Ti oxides of magnetite and ilmenite (Fig. 5.4). Some ilmenites occur as euhedral to

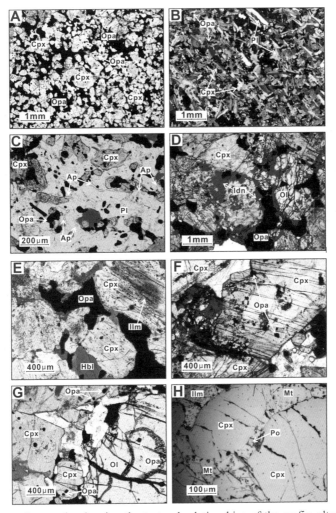

Fig. 5.3 Photomicrographs showing the textural relationships of the mafic–ultramafic rocks in the Wajilitag deposit. (A) Net-texture of equigranular clinopyroxene filled with interstitial Fe–Ti oxides (up to 30 modal%) in the fine-grained pyroxenite, and lots of Fe–Ti oxides also enclosed in the clinopyroxene as inclusions (single polar); (B) Equigranular clinopyroxene and plagioclase with interstitial Fe–Ti oxides in the gabbro (crossed polars); (C) Abundant apatite crystals (up to 5 modal%) in the gabbro (single polar); (D) Olivine in the olivine pyroxenite is partly replaced by iddingsite (single polar); (E) Clinopyroxene rimmed with brown hornblende and two sets of ilmenite lamellae along the prismatic cleavage of clinopyroxene in the coarse-grained pyroxenite (single polar); (F) The zonal structure of clinopyroxene in the gabbro with a pinkish rim (single polar); (G) Euhedral Fe–Ti oxide inclusions enclosed in the olivine of the olivine pyroxenite (single polar); (H) Disseminated pyrrhotite along the boundaries of magnetite in the fine-grained pyroxenite (single polar under reflected light). Ap=apatite, Cpx=clinopyroxene, Hbl=hornblende, Idn=iddingsite, Ilm=ilmenite, Mt=magnetite, Ol=olivine, Opa=opaque minerals, Pl=plagioclase, Po=pyrrhotite

subhedral grains coexisting with magnetite or as commonly oriented exsolution lamellae in the magnetite or clinopyroxene (Fig. 5.4A and Fig. 5.3E). Besides, the fine-grained pyroxenite contains minor pyrrhotites as disseminated grains along the boundaries of Fe–Ti oxides (Fig. 5.3H and Fig. 5.4B).

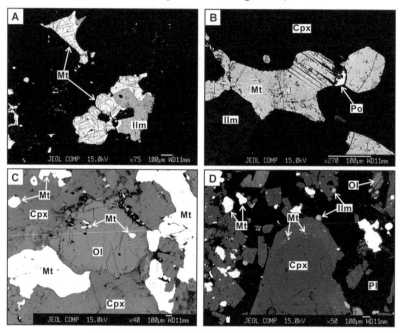

Fig. 5.4 Backscattered electronic (BSE) images of the major minerals from the Wajilitag deposit. (A) Coexisting magnetite and ilmenite, as well as oriented ilmenite exsolution lamellae in the magnetite (Wjl070105); (B) Disseminated pyrrhotite along the boundary of coexisting magnetite–ilmenite in the fine-grained pyroxenite (ZK4001); (C) Magnetite inclusions enclosed in the olivine and clinopyroxene as well as irregular magnetite interstitial to them (Wjl070101); (D) The presence of magnetite grains along the boundary between the core and rim of the zoned clinopyroxene, as well as magnetite and ilmenite grains interstitial to the silicate minerals (Wjl070105). Mineral abbreviations are the same as in Fig. 5.3

5.1.2 Mineral Geochemistry of the Wajilitag Deposit

Four representative samples from the olivine pyroxenite (Wjl070101), coarse-grained pyroxenite (ZK4002), fine-grained pyroxenite (ZK4001) and gabbro with zoned clinopyroxene (Wjl070105) in the Wajilitag deposit were chosen for analyzing their mineral compositions of olivine, clinopyroxene, magnetite and ilmenite by an electron probe micro analyzer (EPMA). Analytical details can be found in Li et al. (2012a), and the results of different minerals are shown in Supplementary Tables II–V, respectively.

In general, the olivine from all of the rock types in the Wajilitag deposit has SiO_2, MgO, FeO and NiO contents of 37.06–39.55, 33.37–38.84, 22.23–29.26 and 0.02–0.20 wt.%, respectively (Supplementary Table II). The Fo values [molar $100\times Mg/(Mg+Fe)$] in the olivine pyroxenite, coarse-grained pyroxenite and gabbro range from 74–76, 67–70 and 67–75 mole%, respectively.

The clinopyroxene has the endmembers of $En_{38-47}Fs_{9-14}Wo_{40-50}$, being Mg-rich augite and diopside (Supplementary Table III). Generally, the clinopyroxene in the olivine pyroxenite and pyroxenite, and the zoned clinopyroxene in the core part in the gabbro show similar chemical compositions (Fig. 5.5). Whereas the zoned clinopyroxene in the rim part in the gabbro is relatively rich in TiO_2 (2.50–3.42 wt.%) and Al_2O_3 (4.45–6.79 wt.%), and low in SiO_2 (46.45–48.56 wt.%) and MgO (12.27–13.64 wt.%). As an exception, two clinopyroxene grains in the coarse-grained pyroxenite have relatively high Na_2O contents (>1 wt.%) with jadeite components of 8%–9% in the clinopyroxene, and this should be retained for further study.

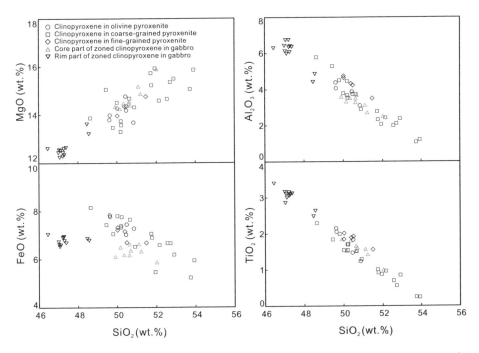

Fig. 5.5 A Harker diagram of SiO_2 versus MgO, Al_2O_3, FeO and TiO_2 for the clinopyroxene in the rocks of the Wajilitag deposit

The magnetites in the intrusion are titanomagnetites with TiO_2, Fe_2O_3 and FeO contents ranging from 1.40–17.86, 30.34–62.08 and 31.34–44.75 wt.%, respectively. They exhibit a relatively high amount of Al_2O_3 of 0.53–4.44 wt.% (mostly >2 wt.%) and variable Cr_2O_3 contents 0–3.07 wt.% (Supplementary Table IV). The ilmenites have TiO_2, Fe_2O_3 and FeO contents ranging from 49.74–54.43, 1.83–6.35 and 34.36–43.45 wt.%, respectively (Supplementary Table V). In general, magnesium and manganese are incompatible in the magnetite and rich in the ilmenite among the coexisting magnetite and ilmenite (Supplementary Table VI).

5.1.3 Crystallization History of Silicate Minerals and Fe–Ti Oxides

Olivine of the olivine pyroxenite rocks has higher Fo values than those of other rocks (Supplementary Table II), being supposed that the former may crystallize much earlier than those in the latter. Nevertheless, the Fo values of all olivines (Fo=67–76 mole%) in the Wajilitag deposit are generally lower than those of olivines (Fo=70–85 mole%) from the Xiaohaizi ultramafic dikes (Jiang et al., 2004; Yang et al., 2007; Zhou et al., 2009). The Ni contents of the olivines studied are also very low (<0.20 wt.%). The magmas parental to the Wajilitag complex thus are considered to have undergone an early removal of olivine with high Fo values (Fo>77 mole%), such as chrysolite, prior to emplacement in the continental crust. In the gabbro, the zonal clinopyroxene grains in the rim parts contain relatively lower MgO, and higher FeO and TiO_2 contents than those in the core parts, indicating the subsolidus equilibration between clinopyroxene and evolved magma during a cooling process.

The Fe–Ti oxides in the Wajilitag deposit are either interstitial to the silicates or enclosed within clinopyroxene, plagioclase and olivine (Figs. 5.3 and 5.4). There is no other silicate mineral in the inclusion-bearing olivine, suggesting that the oxides in the olivine could not be xenocrysts (Pang et al., 2008). Some oxide inclusions have clearly crystal shapes (Fig. 5.3G). Most likely these inclusions were trapped in the growing olivine crystals. It was known that the Fe–Ti oxide crystallization usually occurred at advanced stages (Juster et al., 1989; Toplis and Carroll, 1995; Thy et al., 2006), but it may crystallize before pyroxene and plagioclase crystals (Wang et al., 2008); whereas olivine is commonly deemed to crystallize preferentially in the silicate melt (Bowen, 1922). The presence of oxide crystals hosted in the olivine of olivine pyroxenites (Fig. 5.3G) supposes that they may appear on the liquidus with olivine and clinopyroxene at a relatively early stage (Pang et al., 2008). Both magnetites and ilmenites hosted therein have slightly higher Cr_2O_3 contents than oxides in other silicates or interstitial to them (Supplementary Tables IV and V), which also indicates that they may crystallize much earlier from a more primitive melt (Wang et al., 2008). Therefore, we suggest that the Fe–Ti oxides of the Wajilitag deposit have experienced a two-stage process of crystallization, during which minor Fe–Ti oxides crystallized earlier than the olivine, clinopyroxene and plagioclase during the early stage, whereas most Fe–Ti oxides are interstitial to the silicate minerals crystallized in the late stage when silicate crystals had already cumulated.

5.1.4 Fe–Ti-Rich Parental Magmas of the Wajilitag Deposit

Abundant Fe–Ti oxides in the mafic–ultramafic rocks may be an indicator of Fe–Ti-rich magmas parental to the Wajilitag deposit. The parental magma composition of the intrusion has been recalculated by the olivine–liquid Fe–Mg exchange equilibrium. The FeO/MgO ratios of 1.1–1.6 are yielded using the equation $K_D = (FeO/MgO)_{olivine}/(FeO/MgO)_{melt}$ with an Fe–Mg exchange coefficient (K_D) of 0.3±0.03 (Roeder and Emslie, 1970), suggesting that the Wajilitag magmas may be considerably enriched in FeO contents. Meanwhile, the presence of exsolution lamellae of ilmenite in clinopyroxene (Fig. 5.3E) indicates that the clinopyroxene should crystallize from Ti-rich melts (Zhou et al., 2005). The proposal of Fe–Ti-rich parental magmas is also supported by the bulk chemical compositions of the Wajilitag mafic–ultramafic rocks (Fig. 5.6) and well documented in some layered mafic–ultramafic intrusions from other LIPs around the world (e.g., Zhou et al., 2005; Wang et al., 2008), which may account for the early crystallization of Fe–Ti oxides in the Wajilitag intrusion (Pang et al., 2008).

Detailed field investigation suggested that a Fe- and Ti-rich Wajilitag magma could not be a result of assimilation due to the absence of Fe- and/or Ti-rich rocks, such as a banded iron formation (BIF), in the basement under the Tarim Block. Fractionation of a considerable quantity of anorthosite may cause an enrichment of Fe and Ti in residual magmas like those anorthosite-related Fe–Ti oxide deposits (Duchesne, 1999; Charlier et al., 2006), but anorthosite was not observed in the Wajilitag complex. The magmas parental to the complex were more likely to be from a Fe–Ti-rich magma source. It is notable that the Wajilitag magmas have high Sm/Yb ratios (Fig. 5.7), indicating that they were derived from low-

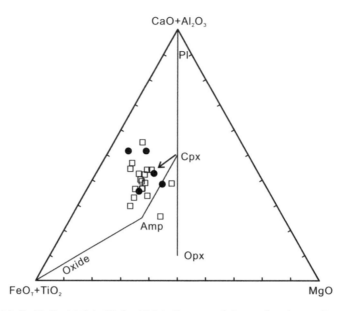

Fig. 5.6 A MgO–(CaO+Al$_2$O$_3$)–(FeO$_T$+TiO$_2$) diagram of the mafic–ultramafic rocks in the Wajilitag deposit, which shows Fe–Ti-rich but poorer in MgO than normal gabbroic rocks as indicated by the arrow. The normal gabbro should follow the line between plagioclase and clinopyroxene (after Zhou et al., 2005). The circle symbols are data from Zhang et al. (2008) and the square symbols are unpublished data from the authors

degree (<10%) partial melts, comparable to the Permian flood basalts in the Tarim Basin (Li et al., 2012b; Yu et al., 2011; Zhang et al., 2010). Such a low-degree partial melting would produce high TiO$_2$ and low Mg# primary magmas (Qin et al., 2011) and cause them to be enriched in Ti and Fe. Besides, layered structures and variable mineral compositions of olivine and clinopyroxene in the deposit suggest that the Wajilitag magmas probably have undergone extensive fractional crystallization derived from a magma chamber. The whole rock geochemistry data of the mafic–ultramafic rocks and syenite from the Wajilitag complex, showing a scattered trend for major oxides and large variations for trace elements (Zhang et al., 2008), also support this proposal. Fractional crystallization of either olivine or clinopyroxene, even plagioclase, could also increase the FeO/MgO ratios and TiO$_2$ contents and cause the magma to be more Fe–Ti-rich (e.g., Barnes and Roeder, 2001; Brooks et al., 1991; Hanski, 1992; Zhou et al., 2005; Wang et al., 2008). Therefore, we suggest that the Fe–Ti-rich Wajilitag intrusive rocks may be derived from a Fe–Ti-rich magma source from a low degree of partial melting and then extensive fractional crystallization of the silicate minerals.

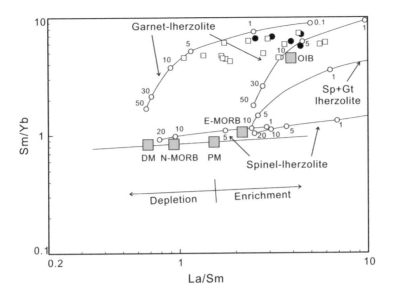

Fig. 5.7 A La/Sm vs. Sm/Yb diagram showing the mafic–ultramafic rocks in the Wajilitag deposit being derived from a low degree (<10%) of partial melting. The lines are non-modal fractional melting curves for garnet lherzolite and spinel lherzolite (after McKenzie and O'Nions, 1991). Numbers on lines refer to percentages of melt. DM=depleted mantle, PM=primitive mantle, N-MORB=normal mid-ocean ridge basalt, Sp=spinel, Gt=garnet. The symbols and data source are the same as in Fig. 5.6

5.1.5 Origin of the Wajilitag Deposit and Its Magmatic Evolution

Mafic–ultramafic intrusions, as a distinct feature of the Emeishan LIP in SW China, such as the Panzhihua, Hongge and Xinjie intrusions, have been widely surveyed and studied (Zhong et al., 2002; Zhou et al., 2005; Wang et al., 2008; Pang et al., 2008). They are considered to be originated from a mantle plume at ~260 Ma (Zhou et al., 2008). The whole rock Sr–Nd isotopic compositions of the mafic–ultramafic rocks in the Wajilitag region exhibit relatively high $\varepsilon_{Nd}(t)$ values (mostly above 2.0) and low initial $^{87}Sr/^{86}Sr$ ratios (0.7035–0.7045) (Jiang et al., 2004; Zhang et al., 2008). They show distinct alkaline affinities and typical intraplate geochemical signatures, consistent with the coeval basalts in the Tarim LIP (Zhang et al., 2010). The igneous rocks in the Wajilitag area were suggested to come from an OIB-like asthenospheric mantle source provided by an upwelling mantle plume in the Early Permian period (Zhang et al., 2008, 2010). We have estimated the liquidus temperature of olivine from the Wajilitag complex using the formula: $T_{liquidus}$ (°C) = 1066 + 12.067×Mg# + 312.3×(Mg#)^2 (Zhou et al., 2009), which is approximately 1253°C (according to the highest Fo value of 76 mole%). Because the olivine with a higher Fo value in the Xiaohaizi ultramafic dikes may have already crystallized out from the parental magma, the liquidus temperature of the primary magma could be much higher, probably over 1300°C (Zhou et al., 2009). Such a high temperature accords with the hypothesis of a mantle plume origin for the Tarim LIP (see Chapter 4). The approximated age of ca. 275 Ma for the Wajilitag complex, although not yet well constrained, is also consistent with the Tarim LIP mantle plume event.

The mantle-derived magmas parental to the Wajilitag deposit must have experienced extensively fractional crystallization and were highly evolved (seen in the above discussion). The crystallization began with olivine and followed by clinopyroxene, while the plagioclase and hornblende may also have been involved. Early crystallized olivine (Fo>77 mole%) might be crystallized during the parental magma passed through the mantle to the crust, and the evolved magma continued its crystallization dominated by clinopyroxene after it emplaced into the middle to upper crust. The evolved magma should be cooled down accompanied by a slowly magmatic crystallization, as indicated by the ilmenite exsolution lamellae in the coexisting magnetite (Fig. 5.4A). A slowly in situ cooling process from the Wajilitag magmas could conduce layered structures to the intrusion due to the influence of gravity, in which the bottom of the intrusion holds mainly olivine and clinopyroxene as the olivine pyroxenite; the pyroxenite with 70–90 modal% clinopyroxene makes up the main body of the intrusion in the middle part; and the gabbro (mainly zoned clinopyroxene and plagioclase) appears on the top in the layered intrusion. The nepheline-bearing syenite should be the latest-stage product of fractional crystallization and represents the Si-rich component of the evolved magma (Li C.-N. et al., 2001; Zhang et al., 2010). That is why they occurred at the uppermost part of the intrusion.

Despite the magnetite or ilmenite inclusions enclosed in the silicate minerals during the early stage, most Fe–Ti oxides crystallized in the late stage as irregular aggregates (Figs. 5.4C and D). The accumulation of disseminated and interstitial magnetite may have been triggered by a Fe_2O_3/FeO ratio of the liquid, a variation in f_{O_2} and pressure, or a volatile content of the silicate magma during fractional crystallization (Zhou et al., 2005; Wang et al., 2008). The ilmenites are probably the result of exsolution from the coexisting magnetite. This process helped remove Mn and Mg from the magnetite and made it more Fe-rich (Fig. 5.8).

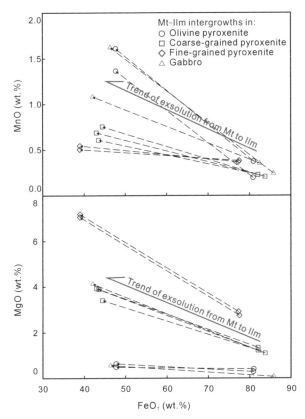

Fig. 5.8 Binary diagrams of FeO_T vs. MnO and MgO for the coexisting magnetite–ilmenite in the mafic–ultramafic rocks of the Wajilitag deposit. The dashed arrows suggest the migration of Mn and Mg from the magnetite to the coexisting ilmenite, indicating a trend of exsolution from magnetite to ilmenite

5.2 Cu–Ni–PGE Mineral Resource Potential in the Tarim LIP

Many LIPs around the world possess world-class Ni–Cu–PGE magmatic ore deposits, such as the Siberian Traps and the Bushveld Complex (Naldrett, 1999, 2004). Although so far no Ni–Cu–PGE deposits have been found in the Tarim LIP, a number of coeval Ni–Cu–PGE magmatic ore deposits, such as the Huangshan–Jing'erquan Cu–Ni sulfide ore belt in eastern Tianshan, the Baishiquan Cu–Ni sulfide deposit in the central Tianshan and the Pobei Cu–Ni sulfide deposit within the Late Paleozoic Beishan Rift, have been reported around the Tarim Basin (Chai et al., 2008; Yang, 2011; Zhou et al., 2004). Potential genetic links between such a widespread intraplate magmatic event and the known Ni–Cu–PGE mineralization have also generated considerable interest among geologists leading to a number of recent investigations (e.g., Mao et al., 2008; Pirajno et al., 2009; Qin et al., 2011; Yuan et al., 2012; Zhang et al., 2011).

PGEs (Os, Ir, Ru, Rh, Pt and Pd) are highly siderophile elements that provide valuable information on the petrogenesis of mantle-derived igneous rocks (e.g., Brügmann et al., 1993; Rehkämper et al., 1999; Momme et al., 2003). The PGE abundances in the CFBs are much lower as compared to lithophile elements, usually at ppb or even ppt level. Nevertheless, they are potential markers of the magmatic process and source nature of the basalts (e.g., Crocket and Paul, 2004; Keays and Lightfoot, 2010; Momme et al., 2002; Qi et al., 2008; Song et al., 2009). They are particularly sensitive due to the extent of their sulfur saturation and sulfide segregation (Barnes et al., 1985; Barnes and Picard, 1993; Keays, 1995; Puchtel and Humayun, 2001). In addition, variations in the concentrations and ratios between different PGEs can provide important information relating to the genesis of magmatic Ni–Cu–PGE sulfide mineralization. A number of studies have proposed that the giant Noril'sk–Talnakh Ni–Cu–PGE sulfide ore is associated with the PGE-depleted basalts of the Nadezhdinsky Formation in the Siberian Traps (Lightfoot and Keays, 2005; Naldrett et al., 1992; Ripley et al., 2003).

In this section, the PGE geochemistry of the Tarim CFBs are systematically studied to address the sulfur saturation and evolution process of the Tarim CFB magmas and to evaluate the Cu–Ni–PGE mineralization potential of the Tarim LIP.

5.2.1 PGE Geochemistry of the Tarim CFBs

To comprehensively know the PGE geochemistry of the Tarim CFBs, 25 basalt samples from different locations in the Tarim LIP were selected to analyze their PGE contents. Among them, 19 samples are Group 1a basalts from the Yingan section of the Keping area (12 samples), the Damusi section of the southwestern Tarim Basin (5 samples) and the Xiahenan section in the central Tarim Basin (2 samples), 2 samples are Group 1b basalts from the Yingan section of the Keping area, and 4 samples are from the SL1, YM5, YM8 and YT6 boreholes in the northern Tarim Uplift (see Fig. 3.1 in Chapter 3 for their locations in the Tarim LIP).

All of the measurements of the PGE contents were conducted at Guangzhou Institute of Geochemistry, Chinese Academy of Sciences. Detailed analytical procedures can be found in Li et al. (2012b), and the results are listed in Table 5.1; the major and trace element abundances as well as their Sr–Nd isotopic compositions (if analyzed) are also given in Supplementary I of Chapter 3.

Table 5.1 PGE and Cu contents of three group basalts in the Tarim LIP

Sample No.	Os	Ir	Ru	Rh	Pt	Pd	Cu	$\sum$PGEs
Yingan section								
Group 1a basalts								
Yg0512-8i	0.026	0.017	0.071	0.020	0.049	0.029	48.5	0.212
Yg0512-8d	0.019	0.011	0.045	0.013	0.043	0.026	45.8	0.157
Yg0512-7f	0.014	0.007	0.041	0.013	0.067	0.084	64.4	0.226
Yg0512-7a	0.015	0.007	0.035	0.011	0.075	0.071	58.5	0.214
Yg0512-6c	0.030	0.019	0.073	0.021	0.077	0.046	53.7	0.266
Yg0512-6b	0.037	0.027	0.097	0.032	0.091	0.085	45.4	0.369
Yg0512-5f	0.022	0.020	0.077	0.031	0.066	0.064	68.7	0.280
Yg0512-5c	0.060	0.050	0.173	0.061	0.149	0.100	41.1	0.593
Yg0512-4k	0.055	0.042	0.148	0.052	0.147	0.124	49.8	0.568
Yg07	0.037	0.020	0.084	0.025	0.069	0.058	74.9	0.293
Yg0512-4a	0.018	0.017	0.053	0.020	0.079	0.065	58.1	0.252
Yg0512-3b	0.046	0.035	0.121	0.039	0.113	0.063	57.9	0.417
Group 1b basalts								
Yg01	0.106	0.067	0.253	0.078	0.129	0.090	52.9	0.723
Yg04	0.056	0.033	0.130	0.034	0.087	0.059	52.7	0.399
Damusi section								
Group 1a basalts								
Txn25-5	0.001	0.000	0.006	0.003	0.066	0.084	223.0	0.160
Txn25-8	0.002	0.002	0.016	0.002	0.219	0.168	103.0	0.409
Txn25-11	0.001	0.002	0.007	0.002	0.031	0.034	106.0	0.078
Txn25-21	0.001	0.004	0.009	0.003	0.089	0.044	68.5	0.150
Txn26-7	0.001	0.004	0.007	0.002	0.126	0.077	64.8	0.217
Xiahenan section								
Group 1a basalts								
XHN13-3	0.002	0.002	0.004	0.009	0.134	0.077	76.8	0.230
XHN14-1	0.003	0.002	0.005	0.005	0.118	0.075	56.2	0.208
North Tarim Uplift								
Group 2 basalts								
SL1-8-10	0.010	0.013	0.025	0.003	0.197	0.041	119	0.288
YM5-22-8	0.001	0.004	0.005	0.001	0.325	0.024	77.2	0.361
YM8-13-7	0.005	0.011	0.012	0.005	0.205	0.112	108	0.349
YT6-9-66	0.001	0.002	0.003	0.001	0.376	0.031	101	0.414

Note: The PGE contents are in the unit of ppb (parts per billion); $\sum$PGEs is the sum of Os, Ir, Ru, Rh, Pt and Pd; Cu contents are also given here for comparison

As shown in Table 5.1, all of the three group basalts from different locations in the Tarim LIP exhibit extremely low PGE contents (Os 0.001–0.106 ppb, Ir 0.000–0.067 ppb, Ru 0.004–0.253 ppb, Rh 0.001–0.078 ppb, Pt 0.031–0.376 ppb, Pd 0.026–0.168 ppb and $\sum$PGEs = 0.078–0.723 ppb). Apart from PGE, the Tarim CFBs are also low in Cu contents (mostly <100 ppm). On the primitive mantle normalized plot, basalts from the same section (or the same location) display a similar PGE pattern (Fig. 5.9). Except those of the Keping basalts, they generally show an enriched trend from Os to Pd in Pt and Pd than the other PGEs, similar to other CFBs around the world (Barnes et al., 1985)

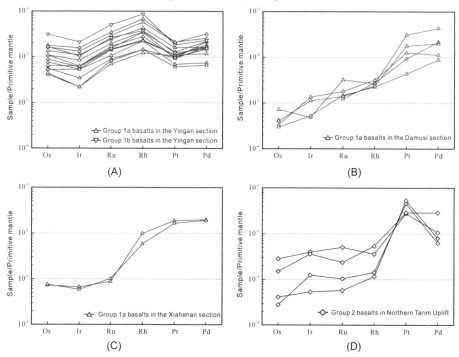

Fig. 5.9 Primitive mantle-normalized PGE patterns of three group basalts from different locations in the Tarim LIP. The primitive mantle normalization values are from McDonough and Sun (1995)

5.2.2 S-Saturated Tarim CFB Magmas

Compared with other well-studied CFBs around the world, the PGE concentrations (represented by Ir and Pd) in the Tarim CFBs are much lower, only similar to those PGE-depleted basalts in the Nadezhdinsky Formation of the Noril'sk region in the Siberian Traps (Table 5.2). Therefore, the Tarim CFBs are strongly depleted in PGE.

Table 5.2 PGE and Cu concentrations of basalts from the Tarim LIP, the Siberian Traps, the Emeishan LIP, the East Greenland CFBs and the Deccan Traps

	N	Ir (ppb)	Pd (ppb)	Cu (ppm)	Pd/Ir	Cu/Pd ($\times 10^5$)
Tarim LIP	36	0–0.067	0.015–0.275	24.7–223	0.34–39.9	1.55–32.9
Siberian Traps (Nadezhdinsky Formation)	15	<0.01	0.07–0.46	20–89	–	0.89–8.86
Siberian Traps (except Nadezhdinsky Formation)	39	0.03–1.68	1.42–17.4	44–498	4.40–276	0.09–0.44
ELIP (high-Ti basalts)	115	0.009–0.88	0.3–32.6	28.5–292	6.11–850	0.03–1.60
Deccan	31	0.05–0.49	2.35–31.8	105–347	12.6–330	0.08–0.68
East Greenland	35	0.05–1.58	3.04–25	43–453	2.07–346	0.09–0.35

Note: N is the number of samples. The data for the Tarim LIP basalts are from this study and Yuan et al. (2012), the data for the Siberian Traps basalts are from Lightfoot and Keays (2005), the data for the Emeishan LIP high-Ti basalts are from Qi and Zhou (2008), Qi et al. (2008) and Song et al. (2009), and the data for the East Greenland basalts are from Momme et al. (2002)

PGEs are strongly partitioned into immiscible sulfide liquids during sulfide segregation because of their extremely high sulfide liquid/silicate melt partition coefficients (Fleet et al., 1991, 1999; Peach et al., 1994), causing their concentrations to decrease in the residual silicate magma. Model calculations indicate that very small amounts of sulfide segregation (as little as 0.57%) can result in significant chalcophile metal depletion, as reported from the Nadezhdinsky Formation in the Noril'sk area (Lightfoot and Keays, 2005). Copper, Palladium and Iridium are often used to judge the sulfur saturation state of a magma (e.g., Momme et al., 2002; Qi et al., 2008 and references therein). Pd and Ir show compatible behavior in the sulfide phase when a suit of lavas undergoes S-saturation differentiation, because their sulfide liquid/silicate melt partition coefficients ($D^{Sul/Sil}$) are the order of 10^3 to 10^5 (Qi et al., 2008; Song et al., 2009). The $D^{Sul/Sil}$ of Cu (10^2 to 10^3) is much lower than the PGEs, which indicates that the Cu/Pd ratio will strongly increase once sulfide liquids segregate from the silicate magma (Bennett et al., 2000; Campbell and Barnes, 1984; Momme et al., 2003). In contrast, both Cu and Pd behave as incompatible elements in S-undersaturated systems, but the Pd/Ir ratio increases during S-undersaturated differentiation (Barnes and Pichard, 1993; Momme et al., 2002). This is because Pd is an incompatible element whereas Ir behaves as a compatible element during typical silicate fractionation of a basaltic magma (Barnes et al.,

1985; Keays, 1995).

The Tarim CFBs have extremely high Cu/Pd ratios ($>10^5$) with a narrow range of lower Pd/Ir ratios (<50), which can be easily distinguished from the S-undersaturated and PGE-undepleted basaltic suites from the Siberian Traps, the Emeishan LIP, the East Greenland CFB and the Deccan Traps (Fig. 5.10A). Besides, the Tarim CFBs also exhibit generally lower Cu/Zr ratios than the PGE-depleted PGE-depleted basalts in the Nadezhdinsky Formation of the Noril'sk region in the Siberian Traps (Fig. 5.10B), indicating a depletion of Cu (Keays and Lightfoot, 2010). These, combined with the extremely low PGE contents as well as the low Cu contents, suggest that the parental magmas of Tarim CFBs were S-saturated before the final eruptions.

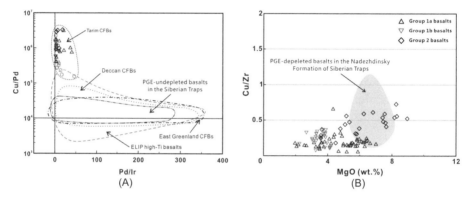

Fig. 5.10 (A) A scattergram of Pd/Ir vs. Cu/Pd showing the fields of the Tarim CFB, the PGE-undepleted basalts in the Siberian Traps, the Emeishan LIP, East Greenland and Deccan. Data sources are the same as for Table 5.1, and the Tarim CFB symbols are the same as in Fig. 5.9, except for the grey hexagon symbols which are data from Yuan et al. (2012). (B) A scattergram of MgO vs. Cu/Zr showing the distributions of the three group basalts in the Tarim LIP. The PGE-depleted Nadezhdinsky Formation basalts in the Siberian Traps (data from Lightfoot and Keays, 2005) are also plotted (shaded area). The source of the data for the Tarim basalts can be found in Supplementary Table I in Chapter 3

5.2.3 A Possible Reason for S-Saturation in the Tarim CFB Magmas

Crustal contamination is commonly considered as the main reason that causes S-saturation in mantle-derived magmas unless there is not a sufficient amount of S in the crustal contaminants (Keays and Lightfoot, 2010; Naldrett, 2004, 2010). The Tarim CFBs in different locations have experienced variable degrees of crustal contamination (see the discussion in Chapter 3), whereas both more contaminated Group 1 basalts and less contaminated Group 2 basalts show a strong depletion in PGE. Moreover, there are poor correlations between the PGE and crustal contamination indicators (e.g., $(Nb/La)_N$; Fig. 5.11). Therefore, crustal contamination during the Tarim basalt eruptions may not trigger S-saturation in their parental magmas. The general CFB-like PGE patterns in most Tarim basalt samples suggest that the S-saturated process probably happened before the fractional crystallization of the magma, and thus are more likely to be related to the region of the magma source.

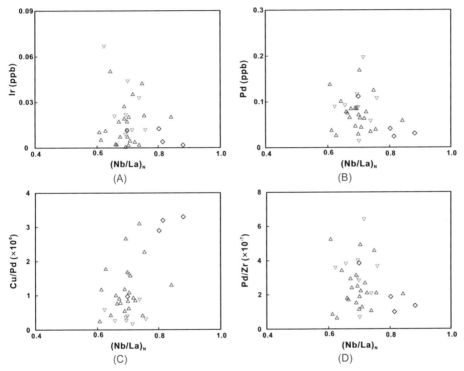

Fig. 5.11 Scattergrams of $(Nb/La)_N$ vs. Ir, Pd, and Cu/Pd, Pd/Zr ratios for the three group basalts in the Tarim LIP. Generally, the PGE displays poor correlations with $(Nb/La)_N$ ratios. Basalt data are from this study and Yuan et al. (2012). Symbols are as in Fig. 5.10B

Mantle-derived magmas will become S-saturated in the source region by low degrees of partial melting, although it could also be S-undersaturated under high f_{O_2} conditions (Jugo et al., 2005; Mungall et al., 2006). Keays (1995) suggested that most mafic magmas produced by less than 25% partial melting are S-saturated assuming that the S capacity of the partial melts is 1000 ppm. Modeling by Naldrett (2010) indicated that the sulfide in the mantle source would entirely dissolve into a silicate liquid when the partial melting degree is up to 11%.

Although the Tarim CFBs in different locations may have variable source compositions, they were derived from a mantle source by low-degree partial melting (e.g., Qin et al., 2011; Yu et al., 2011; Zhang et al., 2010; Zhou et al., 2009). Calculations based on the available trace element geochemistry from the Tarim CFBs suggest that their parent magmas were derived from <5% partial melting of the mantle source (Fig. 5.12), consistent with their alkaline nature and LILE- and LREE-enriched trace element signature (see Chapter 3).

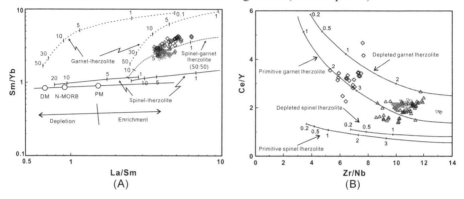

Fig. 5.12 La/Sm vs. Sm/Yb diagram (A) and Zr/Nb vs. Ce/Y diagram (B) showing the melting curves for the Tarim CFBs, indicating that their parent magmas were derived from <5% partial melting. Reference curves are after McKenzie and O'Nions (1991) and Hardarson and Fitton (1991). Symbols are as in Fig. 5.10B

The S content of the mantle reservoir (e.g., primitive mantle, depleted mantle or fertile mantle) is generally about 150–250 ppm (McDonough and Sun, 1995; Lesher and Stone, 1996; Palme and O'Neill, 2003), and therefore melts derived from ca. 5% partial melting would contain 3000 to 5000 ppm S from a columnar melting regime (Keays, 1995). Previous studies suggested that the Tarim CFB magmas originated from a depth equivalent to 17–22 kbar (garnet–spinel transition) and had a mantle plume source temperature of about 1200–1300°C (Zhou et al., 2009; Tian et al., 2010). The sulfur concentration at sulfide saturation (SCSS) for the Tarim CFB magmas can therefore be estimated using the empirical equation provided by Li and Ripley (2009):

$$\ln X_S = -1.76 - 0.474(10^4/T) - 0.021(P) + 5.559 X_{FeO} + 2.565 X_{TiO_2} + 2.709 X_{CaO} - 3.192 X_{SiO_2} - 3.049 X_{H_2O}$$

where T is the temperature in Kelvin, P is the pressure in kbar and X is the mole fraction. As shown in Table 5.3, the calculated SCSS for the parental magmas of the three group basalts at the pressure of 17 kbar and the temperature of 1300°C range from 859 to 1929 ppm, which is much lower than the S content (>3000 ppm) produced by their partial melts.

In this scenario, a substantial amount of residual sulfide would be left behind in the mantle, and only a small portion of the PGEs would be released by the silicate partial melts, which may account for the significant PGE depletion of all the Tarim CFBs. Although the oxidation state in the magma source is yet unknown and needs further investigation, the fact that all the known Tarim CFBs in the Tarim LIP are exclusively PGE-depleted also indicates that their parental magmas have been S-saturated during the same stage, probably when they emerged from the mantle.

Table 5.3 Estimate of SCSS for the parental magmas of three group basalts in the Tarim LIP

Basalt group	SiO_2 (wt.%)	TiO_2 (wt.%)	FeO (wt.%)	CaO (wt.%)	SCSS (ppm)
Group 1a	44.1–47.6	3.5–4.6	9.8–14.9	6.8–9.6	1172–1929
Group 1b	46.1–50.9	2.8–4.2	11.4–14.4	4.3–9.4	1121–1634
Group 2	46.0–52.3	2.0–3.1	7.6–11.1	4.9–11.0	859–1359

Note: CFBs are generally anhydrous so the H_2O contents of the Tarim CFB magmas are probably close to zero. Assuming Fe_2O_3 as 0.15 total iron. Data source can be found in Supplementary Table I in Chapter 3

The low degree partial melting of parental magmas provided seems to account for the strong PGE depletion in all of the Tarim CFBs. Nevertheless, the bell-like PGE pattern of the basalts from the Yingan section, which is quite different from other basalts in the Tarim LIP, implies that the PGE geochemistry of the Keping basalts may also be controlled by other magmatic process.

Fig. 5.13 presents the variations of PGE concentrations, $\varepsilon_{Nd}(t)$ values and $(La/Yb)_N$ ratios for the basalts from the base to the top units of the Yingan section. In the Kaipaizileike sequence, PGE concentrations show a slight increase from the upper Kai1 to lower Kai5, and decrease in the Kai5 unit from the lower part to the upper part. This feature is not likely to have resulted from crustal assimilation because basalts in the Kai5 unit have relatively higher $\varepsilon_{Nd}(t)$ values. Basalts in the upper part of the Kai4 unit and lower part of the Kai5 unit show a marked increase of $\varepsilon_{Nd}(t)$ values from –3.4 to –1.8 and their $^{87}Sr/^{86}Sr_i$ ratios decrease from

0.70691 to 0.70614 (see Supplementary Table I in Chapter 3). More likely, it may indicate a continuous influx of relatively primitive and uncontaminated magma of the same lineage into a chamber occupied by evolved residual magma.

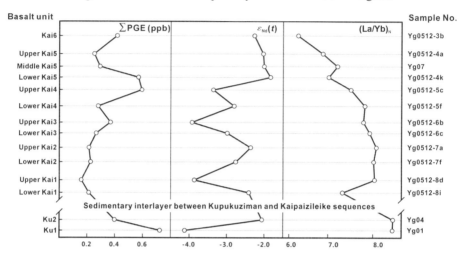

Fig. 5.13 PGE, $\varepsilon_{Nd}(t)$ and $(La/Yb)_N$ systematic variations for the basalts along the whole Yingan section, in the Keping area. On the left are the corresponding positions of the sample on the right; for example, lower Kai1 represents the lower part of the first basalt unit in the Kaipaizileike sequence. Nd isotope and rare earth element (La and Yb) data can be found in the Supplementary Table I in Chapter 3

Mixing of an evolved magma with appropriate amounts of a primitive magma is capable of achieving sulfur saturation in the hybrid (Li C.-S. et al., 2001), which may have triggered a certain extent of sulfide segregation and PGE depletion in the Kaipaizileike sequence, although the feature is not so prominent due to their extremely low PGE contents. Such a magma chamber replenishing process is also supported by the general decreasing trend of the $(La/Yb)_N$ ratios from the base to the top basaltic units in the Yingan section, which eventually led to multiple basaltic flows in the Keping area (Yu et al., 2010).

5.2.4 Implications for the Cu–Ni–PGE Mineral Resource Potential in the Tarim LIP

The above considerations indicate that the degree of partial melting plays an important role in the Cu–Ni–PGE mineralization in the Tarim LIP. Due to very low degrees of partial melting, the Tarim CFBs are extremely PGE-depleted, which may account for the absence of Cu–Ni–PGE sulfide deposits in the Tarim LIP. Mafic–ultramafic rocks formed by a high-degree of partial melting from the mantle source in the Tarim LIP should receive more attention in terms of their Cu–Ni–PGE mineralization potential. In fact, a number of similar deposits have been reported in the eastern Tianshan and Beishan area, and recent studies suggest that they may also be a part of the Tarim LIP (Pirajno et al., 2009; Qin et al., 2011; Su et al., 2011). It is also noted that the PGE geochemistry of basalts in the Yingan section indicates that magma replenishment in the magma chamber of the Keping basalts may trigger S-saturation in the evolved magma. The latter may cause the magma to become S-saturated again and to segregate the saturated sulfide in the crust before final eruptions (Naldrett, 2004), which may have been possible to develop a Cu–Ni–PGE sulfide deposit in the Keping area.

References

Barnes, S.J., Picard, C.P., 1993. The behaviour of platinum-group elements during partial melting, crystal fractionation, and sulphide segregation: An example from the Cape Smith Fold Belt, northern Quebec. Geochimica et Cosmochimica Acta, 57(1): 79-87.

Barnes, S.J., Roeder, P.L., 2001. The range of spinel compositions in terrestrial mafic and ultramafic rocks. Journal of Petrology, 42: 2279-2302.

Barnes, S.J., Naldrett, A.J., Gorton, M.P., 1985. The origin of the fractionation of platinum-group elements in terrestrial magmas. Chemical Geology, 53(3-4): 303-323.

Bennett, V.C., Norman, M.D., Garcia, M.O., 2000. Rhenium and platinum group element abundances correlated with mantle source components in Hawaiian picrites: Sulphides in the plume. Earth and Planetary Science Letters, 183(3-4): 513-526.

Bowen, N.L., 1922. The reaction principle in petrogenesis. Journal of Geology, 30: 177-198.

Brooks, C.K., Larsen, I.M., Nielsen, T.F.D., 1991. Importance of iron-rich tholeiitic magmas at divergent plate margins — A reappraisal. Geology, 19: 269-272.

Brügmann, G.E., Naldrett, A.J., Asif, M., Lightfoot, P.C., Gorbachev, N.S., Fedorenko, V.A., 1993. Siderophile and chalcophile metals as tracers of the evolution of the Siberian Trap in the Noril'sk region, Russia. Geochimica et Cosmochimica Acta, 57(9): 2001-2018.

Bryan, S.E., Ernst, R.E., 2008. Revised definition of Large Igneous Provinces (LIPs). Earth-Science Reviews, 86(1-4): 175-202.

Chai, F.-M., Zhang, Z.-C., Mao, J.-W., Dong, L.-H., Zhang, Z.-H., Wu, H., 2008. Geology, petrology and geochemistry of the Baishiquan Ni–Cu-bearing mafic–ultramafic intrusions in Xinjiang, NW China: Implications for tectonics and genesis of ores. Journal of Asian Earth Sciences, 32(2-4): 218-235.

Campbell, I.H., Barnes, S.J., 1984. A model for the geochemistry of the platinum-group elements in magmatic sulfide deposits. Canadian Mineralogist, 22(1): 151-160.

Charlier, B., Duchesne, J.C., Vander Auwera, J., 2006. Magma chamber processes in the Tellnes ilmenite deposit (Rogaland Anorthosite Province, SW Norway) and the formation of Fe–Ti ores in massif-type anorthosites. Chemical Geology, 234: 264-290.

Crocket, J.H., Paul, D.K., 2004. Platinum-group elements in Deccan mafic rocks: A comparison of suites differentiated by Ir content. Chemical Geology, 208(1-4): 273-291.

Duchesne, J.C., 1999. Fe–Ti deposits in Rogaland anorthosites (South Norway): Geochemical characteristics and problems of interpretation. Mineralium Deposita, 34: 182-198.

Fleet, M.E., Tronnes, R.G., Stone, W.E., 1991. Partitioning of platinum group elements in the Fe–O–S system to 11 GPa and their fractionation in the mantle and meteorites. Journal of Geophysical Research, 96(B13): 21949-21958.

Fleet, M.E., Crocket, J.H., Liu, M.-H., Stone, W.E., 1999. Laboratory partitioning of platinum-group elements (PGE) and gold with application to magmatic sulfide–PGE deposits. Lithos, 47(1-2): 127-142.

Gao, Y.-S., 2007. Geological characteristics of Wajilitag vanadic titanomagnetite deposit and its prospecting recommendations. Xinjiang Iron and Steel, 102: 8-9 (in Chinese).

Hardarson, B.S., Fitton, J.G., 1991. Increased mantle melting beneath Snaefellsjokull volcano during late Pleistocene glaciation. Nature, 353: 62-64.

Hanski, E.J., 1992. Petrology of the Pechenga ferropicrites and cogenetic, Ni-bearing gabbro–wehrlite intrusions, Kola Peninsula, Russia. Geological Survey of Finland Bulletin, 367: 192.

Jugo, P.J., Luth, R.W., Richards, J.P., 2005. Experimental data on the speciation of sulfur as a function of oxygen fugacity in basaltic melts. Geochim. Cosmochim. Acta, 69(2): 497-503.

Jiang, C.-Y., Jia, C.-Z., Li, L.-C., Zhang, P.-B., Lu, D.-R., Bai, K.-Y., 2004. Source of the Fe-enriched-type high-Mg magma in Mazhartag region, Xinjiang. Acta Geologica Sinica, 78: 770-780 (in Chinese with English abstract).

Juster, T.C., Grove, T.L., Perfit, M.R., 1989. Experimental constraints on the generation of Fe–Ti basalts, andesites, and rhyodacites at the Galapagos Spreading Centre, 85°W and 95°W. Journal of Geophysical Research, 94: 9251-9274.

Keays, R.R., 1995. The role of komatiitic and picritic magmatism and S-saturation in the formation of ore deposits. Lithos, 34(1-3): 1-18.

Keays, R.R., Lightfoot, P.C., 2010. Crustal sulfur is required to form magmatic Ni–Cu sulfide deposits: Evidence from chalcophile element signatures of Siberian and Deccan Trap basalts. Mineralium Deposita, 45(3): 241-257.

Lesher, C.M., Stone, W.E., 1996. Exploration geochemistry of komatiites. In: Wyman, D. (Ed.), Igneous Trace Elements Geochemistry: Applications for Massive Sulphide Exploration. Geological Society of Canada Short Course Notes, pp. 153-204.

Li, C.-N., Lu, F.-X., Chen, M.-H., 2001. Research on petrology of the Wajilitag complex body in north edge in the Tarim Basin. Xinjiang Geology, 19: 38-43 (in Chinese with English abstract).

Li, C.-S., Maier, W.D., de Waal, S.A., 2001. The role of magma mixing in the genesis of PGE mineralization in the Bushveld Complex: Thermodynamic calculations and new interpretations. Economic Geology, 96(3): 653-662.

Li, C.-S., Ripley, E.M., 2009. Sulfur contents at sulfide–liquid or anhydrite saturation in silicate melts: Empirical equations and example applications. Economic Geology, 104(3): 405-412.

Li, Y.-Q., Li, Z.-L., Chen, H.-L., Yang, S.-F., Yu, X., 2012a. Mineral characteristics and metallogenesis of the Wajilitag layered mafic–ultramafic intrusion and associated Fe–Ti–V oxide deposit in the Tarim Large Igneous Province, northwest China. Journal of Asian Earth Sciences, 49: 161-174.

Li, Y.-Q., Li, Z.-L., Sun, Y.-L., Santosh, M., Langmuir, C.H., Chen, H.-L., Yang, S.-F., Chen, Z.-X., Yu, X., 2012b. Platinum-group elements and geochemical characteristics of the Permian continental flood basalts in the Tarim Basin, Northwest China: Implications for the evolution of the Tarim Large Igneous Province. Chemical Geology, 328: 278-289.

Lightfoot, P.C., Keays, R.R., 2005. Siderophile and chalcophile metal variations in flood basalts from the Siberian Trap, Noril'sk Region: Implications for the origin of the Ni–Cu–PGE sulfide ores. Economic Geology, 100(3): 439-462.

Lu, F.-X., Li, C.-N., Chen, M.-H., 2000. Study on alkali rock belt and metallogenic geological conditions of rare earth elements, Gem and diamond in the north Tarim Basin. Xinjiang National 305 Project Office Open File, pp. 75-110 (in Chinese).

McKenzie, D., O'Nions, R.K., 1991. Partial melt distributions from inversion of rare earth element concentrations. Journal of Petrology, 32(5): 1021-1091.

McDonough, W.F., Sun, S.-S., 1995. The composition of the Earth. Chemical Geology, 120(3-4): 223-253.

Mungall, J.E., Hanley, J.J., Arndt, N.T., Debecdelievre, A., 2006. Evidence from meimechites and other low-degree mantle melts for redox controls on mantle-crust fractionation of platinum-group elements. Proc. Natl. Acad. Sci. U.S.A., 103(34): 12695-12700.

Mao, J.-W., Pirajno, F., Zhang, Z.-H., Chai, F.-M., Wu, H., Chen, S.-P., Cheng, L.-S., Yang, J.-M., Zhang, C.-Q., 2008. A review of Cu–Ni sulphide deposits in the Chinese Tianshan and Altay orogens (Xinjiang Autonomous Region, NW China): Principal characteristics and ore-forming processes. Journal of Asian Earth Sciences, 32(2-4): 184-203.

Momme, P., Tegner, C., Brooks, C.K., Keays, R.R., 2002. The behaviour of platinum-group elements in basalts from the East Greenland rifted margin. Contributions to Mineralogy and Petrology, 143(2): 133-153.

Momme, P., Óskarsson, N., Keays, R.R., 2003. Platinum-group elements in the Icelandic rift system: Melting processes and mantle sources beneath Iceland. Chemical Geology, 196(1-4): 209-234.

Naldrett, A.J., 1999. World-class Ni–Cu–PGE deposits: Key factors in their genesis. Mineralium Deposita, 34: 227-240.

Naldrett, A.J., 2004. Magmatic Sulfide Deposits: Geology, Geochemistry and Exploration.

Berlin: Springer, p. 728.

Naldrett, A.J., 2010. From the mantle to the bank: The life of a Ni–Cu–(PGE) sulfide deposit. South African Journal of Geology, 113(1): 1-32.

Naldrett, A.J., Lightfoot, P.C., Fedorenko, V.A., Doherty, W., Gorbachev, N.S., 1992. Geology and geochemistry of intrusions and flood basalts of the Noril'sk region, USSR, with implications for the origin of the Ni–Cu ores. Economic Geology, 87(4): 975-1004.

Palme, H., O'Neill, H.St.C., 2003. Cosmochemical estimates of mantle composition. In: Carlson, R.W. (Ed.), Treatise on Geochemistry: The Mantle and Core, vol. 2. Elsevier, pp. 1-38.

Pang, K.-N., Li, C.-S., Zhou, M.-F., Ripley, E.M., 2008. Abundant Fe–Ti oxide inclusions in olivine from the Panzhihua and Hongge layered intrusions, SW China: Evidence for early saturation of Fe–Ti oxides in ferrobasaltic magma. Contributions to Mineralogy and Petrology, 156: 307-321.

Peach, C.L., Mathez, E.A., Keays, R.R., Reeves, S.J., 1994. Experimentally determined sulfide melt–silicate melt partition coefficients for iridium and palladium. Chem. Geol., 117(1-4): 361-377.

Pirajno, F., 2013. The Geology and Tectonic Settings of China's Mineral Deposits. Dordrecht: Springer, p. 679.

Pirajno, F., Ernst, R.E., Borisenko, A.S., Fedoseev, G., Naumov, E.A., 2009. Intraplate magmatism in Central Asia and China and associated metallogeny. Ore Geology Reviews, 35(2): 114-136.

Puchtel, I.S., Humayun, M., 2001. Platinum group element fractionation in a komatiitic basalt lava lake. Geochimica et Cosmochimica Acta, 65(17): 2979-2993.

Qi, L., Zhou, M.-F., 2008. Platinum-group elemental and Sr–Nd–Os isotopic geochemistry of Permian Emeishan flood basalts in Guizhou Province, SW China. Chemical Geology, 248(1-2): 83-103.

Qi, L., Wang, C.Y., Zhou, M.-F., 2008. Controls on the PGE distribution of Permian Emeishan alkaline and peralkaline volcanic rocks in Longzhoushan, Sichuan Province, SW China. Lithos, 106(3-4): 222-236.

Qin, K.-Z., Su, B.-X., Sakyi, P.A., Tang, D.-M., Li, X.-H., Sun, H., Xiao, Q.-H., Liu, P.-P., 2011. SIMS zircon U–Pb geochronology and Sr–Nd isotopes of Ni–Cu-bearing mafic–ultramafic intrusions in Eastern Tianshan and Beishan in correlation with flood basalts in Tarim Basin (NW China): Constraints on a ca. 280 Ma mantle plume. American Journal of Science, 311: 237-260.

Rehkämper, M., Halliday, A.N., Fitton, J.G., Lee, D.-C., Wieneke, M., Arndt, N.T., 1999. Ir, Ru, Pt and Pd in basalts and komatiites: New constraints for the geochemical behavior of the platinum-group elements in the mantle. Geochimica et Cosmochimica Acta, 63(22): 3915-3934.

Ripley, E.M., Lightfoot, P.C., Li, C.-S., Elswick, E.R., 2003. Sulfur isotopic studies of continental flood basalts in the Noril'sk region: Implications for the association between lavas and ore-bearing intrusions. Geochimica et Cosmochimica Acta, 67(15): 2805-2817.

Roeder, P.L., Emslie, R.F., 1970. Olivine–liquid equilibrium. Contributions to Mineralogy and Petrology, 29: 275-289.

Song, X.-Y., Zhou, M.-F., Tao, Y., Xiao, J.-F., 2008. Controls on the metal compositions of magmatic sulfide deposits in the Emeishan large igneous province, SW China. Chemical Geology, 253(1-2): 38-49.

Song, X.-Y., Keays, R.R., Xiao, L., Qi, H.-W., Ihlenfeld, C., 2009. Platinum-group element geochemistry of the continental flood basalts in the central Emeishan Large Igneous Province, SW China. Chemical Geology, 262(3-4): 246-261.

Su, B.-X., Qin, K.-Z., Sakyi, P.A., Li, X.-H., Yang, Y.-H., Sun, H., Tang, D.-M., Liu, P.-P., Xiao, Q.-H., Malaviarachchi, S.P.K., 2011. U–Pb ages and Hf–O isotopes of zircons from Late Paleozoic mafic–ultramafic units in the southern Central Asian Orogenic Belt: Tectonic implications and evidence for an Early Permian mantle plume. Gondwana Research, 20(2-3): 516-531.

Thy, P., Lesher, C.E., Nielsen, T.F.D., Brooks, C.K., 2006. Experimental constraints on the Skaergaard liquid line of descent. Lithos, 92: 154-180.

Tian, W., Campbell, I.H., Allen, C.M., Guan, P., Pan, W.-Q., Chen, M.-M., Yu, H.-J., Zhu, W.-P., 2010. The Tarim picrite–basalt–rhyolite suite, a Permian flood basalt from Northwest China with contrasting rhyolites produced by fractional crystallization and anatexis. Contributions to Mineralogy and Petrology, 160(3): 407-425.

Toplis, M.J., Carroll, M.R., 1995. An experimental study of the influence of oxygen fugacity on Fe–Ti oxide stability, phase relations, and mineral–melt equilibria in ferro-basaltic systems. Journal of Petrology, 36: 1137-1170.

Wang, C.Y., Zhou, M.-F., Zhao, D., 2008. Fe–Ti–Cr oxides from the Permian Xinjie mafic–ultramafic layered intrusion in the Emeishan Large Igneous Province, SW China: Crystallization from Fe- and Ti-rich basaltic magmas. Lithos, 102: 198-217.

Yang, S.-F., Chen, H.-L., Ji, D.-W., Li, Z.-L., Dong, C.-W., Jia, C.-Z., Wei, G.-Q., 2005. Geological process of early to middle Permian magmatism in Tarim Basin and its geodynamic significance. Geological Journal of China Universities, 11: 504-511 (in Chinese with English abstract).

Yang, S.-F., Yu, X., Chen, H.-L., Li, Z.-L., Wang, Q.-H., Luo, J.-C., 2007. Geochemical characteristics and petrogenesis of Permian Xiaohaizi ultrabasic dike in Bachu area, Tarim Basin. Acta Petrologica Sinica, 23: 1087-1096 (in Chinese with English abstract).

Yang, S.-H., 2011. The Permian Pobei Mafic–Ultramafic Intrusion (NE Tarim, NW China) and Associated Sulfide Mineralization. Ph. D. Thesis, The University of Hong Kong, p. 261.

Yuan, F., Zhou, T.-F., Zhang, D.-Y., Jowitt, S.M., Keays, R.R., Liu, S., Fan, Y., 2012. Siderophile and chalcophile metal variations in basalts: Implications for the sulfide saturation history and Ni–Cu–PGE mineralization potential of the Tarim continental flood basalt province, Xinjiang, China. Ore Geology Reviews, 45: 5-15.

Yu, X., Chen, H.-L., Yang, S.-F., Li, Z.-L., Wang, Q.-H., Li, Z.-H., 2010. Distribution characters of Permian basalts and their geological significance in the Kalpin area, Xinjiang. Journal of Stratigraphy, 34(2): 127-134 (in Chinese with English abstract).

Yu, X., Yang, S.-F., Chen, H.-L., Chen, Z.-Q., Li, Z.-L., Batt, G.E., Li, Y.-Q., 2011. Permian flood basalts from the Tarim Basin, Northwest China: SHRIMP zircon U–Pb dating and geochemical characteristics. Gondwana Research, 20(2-3): 485-497.

Zhang, C.-L., Xu, Y.-G., Li, Z.-X., Wang, H.-Y., Ye, H.-M., 2010. Diverse Permian magmatism in the Tarim Block, NW China: Genetically linked to the Permian Tarim mantle plume. Lithos, 119: 537-552.

Zhang, C.-L., Li, X.-H., Li, Z.-X., Ye, H.-M., Li, C.-N., 2008. A Permian layered intrusive complex in the western Tarim Block, Northwestern China: product of a ca. 275 Ma mantle plume. Journal of Geology, 116: 269-287.

Zhang, M.-J., Li, C.-S., Fu, P.-E., Hu, P.-Q., Ripley, E.M., 2011. The Permian Huangshanxi Cu–Ni deposit in western China: intrusive–extrusive association, ore genesis, and exploration implications. Mineralium Deposita, 46(2): 153-170.

Zhong, H., Zhou, X.-H., Zhou, M.-F., Sun, M., Liu, B.-G., 2002. Platinum-group element geochemistry of the Hongge layered intrusion in the Pan-Xi area, southwestern China. Mineralium Deposita, 37: 226–239.

Zhou, M.-F., Lesher, C.M., Yang, Z.-X., Li, J.-W., Sun, M., 2004. Geochemistry and petrogenesis of 270 Ma Ni–Cu–(PGE) sulfide-bearing mafic intrusions in the Huangshan district, eastern Xinjiang, Northwest China: Implications for the tectonic evolution of the Central Asian orogenic belt. Chemical Geology, 209(3-4): 233-257.

Zhou, M.-F., Robinson, P.T., Lesher, C.M., Keays, R.R., Zhang, C.-J., Malpas, J., 2005. Geochemistry, petrogenesis, and metallogenesis of the Panzhihua gabbroic layered intrusion and associated Fe–Ti–V oxide deposits, Sichuan Province, SW China. Journal of Petrology, 46: 2253-2280.

Zhou, M.-F., Arndt, N.T., Malpas, J., Wang, C.Y., Kennedy, A.K., 2008. Two magma series and associated ore deposits types in the Permian Emeishan Large Igneous Province, SW China. Lithos, 103: 352-368.

Zhou, M.-F., Zhao, J.-H., Jiang, C.-Y., Gao, J.-F., Wang, W., Yang, S.-H., 2009. OIB-like, heterogeneous mantle sources of Permian basaltic magmatism in the western Tarim Basin, NW China: Implications for a possible Permian large igneous province. Lithos, 113: 583-594.

Supplementary Table II Chemical compositions of olivine in the Wajilitag deposit

Sample No.	Wjl070101 (olivine pyroxenite)																			
Mineral No.	Ol-1					Ol-2					Ol-3					Ol-4				
Analysis No.	1	2	3	4	5	1	2	3	4	5	1	2	3	4	5	1	2	3	4	5
SiO$_2$ (wt.%)	38.42	38.30	38.79	38.80	38.44	38.40	38.52	38.53	38.66	38.40	38.31	38.72	38.61	38.53	38.95	38.58	38.42	39.05	38.44	38.79
FeO	23.71	24.03	22.92	23.24	23.22	23.61	23.42	23.76	23.07	23.93	23.36	23.91	23.10	23.86	22.81	23.09	24.01	22.23	23.10	22.32
MnO	0.38	0.36	0.39	0.31	0.37	0.39	0.42	0.37	0.32	0.39	0.38	0.28	0.35	0.37	0.33	0.40	0.37	0.33	0.43	0.42
MgO	37.88	37.84	37.66	37.53	37.48	37.83	37.60	37.32	37.72	37.68	38.22	37.46	37.72	37.74	38.17	38.26	37.76	38.11	38.04	38.22
NiO	0.09	0.12	0.20	0.07	0.09	0.09	0.07	0.17	0.04	0.10	0.08	0.10	0.16	0.13	0.13	0.17	0.04	0.13	0.12	0.07
Total	100.5	100.6	99.96	99.95	99.61	100.3	100.0	100.2	99.82	100.5	100.3	100.5	99.93	100.6	100.4	100.5	100.6	99.85	100.1	99.82
Si (O = 4)	1.00	1.00	1.01	1.01	1.01	1.00	1.01	1.01	1.01	1.00	1.00	1.01	1.01	1.00	1.01	1.00	1.00	1.02	1.00	1.01
Fe^{2+}	0.52	0.52	0.50	0.51	0.51	0.52	0.51	0.52	0.50	0.52	0.51	0.52	0.50	0.52	0.49	0.50	0.52	0.48	0.50	0.49
Mn	0.01	0.01	0.01	0.01	0.01	0.01	0.01	0.01	0.01	0.01	0.01	0.01	0.01	0.01	0.01	0.01	0.01	0.01	0.01	0.01
Mg	1.47	1.47	1.46	1.46	1.46	1.47	1.46	1.45	1.47	1.46	1.48	1.45	1.47	1.46	1.47	1.48	1.47	1.48	1.48	1.48
Ni	0.00	0.00	0.00	0.00	0.00	0.00	0.00	0.00	0.00	0.00	0.00	0.00	0.00	0.00	0.00	0.00	0.00	0.00	0.00	0.00
∑cation	3.00	3.00	2.99	2.99	2.99	3.00	2.99	2.99	2.99	3.00	3.00	2.99	2.99	3.00	2.99	3.00	3.00	2.98	3.00	2.99
Fo (mol%)	74.00	73.73	74.54	74.21	74.20	74.06	74.10	73.67	74.45	73.72	74.46	73.63	74.42	73.81	74.89	74.70	73.70	75.33	74.58	75.32

(To be continued)

(Supplementary Table II)

Sample No.	ZK4002 (coarse-grained pyroxenite)											Wjl070105 (gabbro)				
Mineral No.	Ol-1					Ol-2						Ol-1				
Analysis No.	1	2	3	4	1	2	3	4	1	2	3	4	1	2	3
SiO_2 (wt.%)	37.33	37.82	36.84	37.30	37.34	37.63	37.16	36.69	38.42	37.75	37.43
FeO	27.33	27.22	28.15	27.15	28.10	28.37	28.35	28.97	22.91	26.39	29.17
MnO	0.43	0.49	0.46	0.49	0.49	0.53	0.47	0.49	0.32	0.56	0.49
MgO	34.52	35.15	34.62	34.60	33.79	33.96	33.94	33.75	38.44	35.03	33.37
NiO	0.02	0.02	0.07	0.11	0.09	0.02	0.05	0.06	–	–	–
Total	99.63	100.7	100.1	99.65	99.81	100.5	99.96	99.96	100.1	99.73	100.5
Si (O = 4)	1.00	1.00	0.99	1.00	1.00	1.00	1.00	0.99	1.00	1.01	1.00
Fe^{2+}	0.61	0.60	0.63	0.61	0.63	0.63	0.64	0.65	0.50	0.59	0.65
Mn	0.01	0.01	0.01	0.01	0.01	0.01	0.01	0.01	0.01	0.01	0.01
Mg	1.38	1.39	1.38	1.38	1.35	1.35	1.36	1.36	1.49	1.39	1.33
Ni	0.00	0.00	0.00	0.00	0.00	0.00	0.00	0.00	–	–	–
Σcation	3.00	3.00	3.01	3.00	3.00	3.00	3.00	3.01	3.00	2.99	3.00
Fo (mol%)	69.23	69.70	68.66	69.42	68.18	68.08	68.08	67.49	74.94	70.28	67.09

Fo = molar 100 × Mg/(Mg + Fe); "–" means not analyzed; no olivine in fine-grained pyroxenite.

Supplementary Table III Chemical compositions of clinopyroxene in the Wajilitag deposit

Sample No. Wjl070105 (gabbro)

Mineral No.	Cpx-1							Cpx-2							Cpx-3						
Analysis No.	1	2	3	4	5	6	7	1	2	3	4	5	6	7	1	2	3	4	5	6	7
Position	Rim		Core			Rim		Rim		Core			Rim		Rim		Core			Rim	
SiO_2 (wt.%)	47.23	46.45	50.70	50.36	49.91	47.30	47.04	47.39	48.46	51.23	51.11	52.03	48.56	47.11	47.22	47.27	50.63	50.65	50.17	47.00	47.07
TiO_2	3.07	3.42	1.60	1.79	1.83	3.11	2.88	3.14	2.50	1.41	1.56	0.98	2.65	3.02	3.18	3.10	1.65	1.50	1.51	3.18	3.13
Al_2O_3	6.39	6.33	3.66	3.49	3.55	6.46	6.15	6.40	4.45	2.68	3.10	2.49	4.89	6.02	6.79	6.11	3.26	3.47	3.29	6.46	6.78
Cr_2O_3	0.03	0.05	0.11	0.12	0.14	0.07	0.00	0.00	0.03	0.15	0.20	0.41	0.00	0.00	0.06	0.06	0.14	0.01	0.37	0.04	0.05
FeO	6.92	7.06	6.34	6.15	6.08	6.80	6.64	6.71	6.89	6.30	6.61	5.84	6.82	6.62	6.94	6.98	6.14	6.59	6.47	6.76	6.55
MnO	0.13	0.13	0.03	0.11	0.12	0.16	0.10	0.08	0.17	0.08	0.08	0.10	0.11	0.15	0.11	0.17	0.08	0.17	0.17	0.17	0.12
MgO	12.46	12.63	14.37	14.46	14.32	12.37	12.58	12.67	13.64	14.85	15.16	15.87	13.24	12.56	12.31	12.63	14.46	14.54	14.22	12.46	12.27
CaO	22.28	22.51	22.49	22.41	22.68	22.34	22.47	22.36	22.66	22.06	21.73	21.57	22.63	22.33	22.15	22.45	22.40	21.99	22.32	22.53	22.44
Na_2O	0.68	0.52	0.41	0.44	0.45	0.90	0.56	0.64	0.45	0.42	0.43	0.37	0.48	0.72	0.77	0.60	0.44	0.47	0.57	0.65	0.88
K_2O	0.00	0.01	0.01	0.00	0.01	0.00	0.00	0.00	0.01	0.00	0.00	0.00	0.00	0.00	0.00	0.00	0.01	0.01	0.00	0.00	0.00
Total	99.19	99.11	99.70	99.33	99.08	99.51	98.41	99.39	99.26	99.18	99.97	99.66	99.38	98.53	99.53	99.36	99.21	99.39	99.10	99.25	99.30
Si (O=6)	1.78	1.76	1.88	1.88	1.87	1.78	1.78	1.78	1.82	1.91	1.89	1.92	1.82	1.78	1.77	1.78	1.89	1.89	1.88	1.77	1.77
Ti	0.09	0.10	0.04	0.05	0.05	0.09	0.08	0.09	0.07	0.04	0.04	0.03	0.07	0.09	0.09	0.09	0.05	0.04	0.04	0.09	0.09

(To be continued)

(Supplementary Table III)

Sample No.: Wjl070105 (gabbro)

Mineral No.	Cpx-1							Cpx-2							Cpx-3						
Analysis No.	1	2	3	4	5	6	7	1	2	3	4	5	6	7	1	2	3	4	5	6	7
Position	Rim		Core			Rim		Rim		Core			Rim		Rim		Core			Rim	
Al	0.28	0.28	0.16	0.15	0.16	0.29	0.27	0.28	0.20	0.12	0.14	0.11	0.22	0.27	0.30	0.27	0.14	0.15	0.15	0.29	0.30
Cr	0.00	0.00	0.00	0.00	0.00	0.00	0.00	0.00	0.00	0.00	0.01	0.01	0.00	0.00	0.00	0.00	0.00	0.00	0.01	0.00	0.00
Fe^{2+}	0.22	0.22	0.20	0.19	0.19	0.21	0.21	0.21	0.22	0.20	0.20	0.18	0.21	0.21	0.22	0.22	0.19	0.21	0.20	0.21	0.21
Mn	0.00	0.00	0.00	0.00	0.00	0.01	0.00	0.00	0.01	0.00	0.00	0.00	0.00	0.00	0.00	0.01	0.00	0.01	0.01	0.01	0.00
Mg	0.70	0.71	0.79	0.80	0.80	0.69	0.71	0.71	0.76	0.82	0.84	0.87	0.74	0.71	0.69	0.71	0.80	0.81	0.79	0.70	0.69
Ca	0.90	0.91	0.89	0.90	0.91	0.90	0.91	0.90	0.91	0.88	0.86	0.85	0.91	0.91	0.89	0.90	0.89	0.88	0.90	0.91	0.90
Na	0.05	0.04	0.03	0.03	0.03	0.07	0.04	0.05	0.03	0.03	0.03	0.03	0.03	0.05	0.06	0.04	0.03	0.03	0.04	0.05	0.06
K	0.00	0.00	0.00	0.00	0.00	0.00	0.00	0.00	0.00	0.00	0.00	0.00	0.00	0.00	0.00	0.00	0.00	0.00	0.00	0.00	0.00
Σcation	4.02	4.02	4.01	4.01	4.02	4.03	4.02	4.02	4.02	4.01	4.01	4.00	4.01	4.02	4.02	4.02	4.01	4.01	4.02	4.02	4.02
Mg#	76.25	76.11	80.17	80.73	80.75	76.41	77.15	77.09	77.92	80.77	80.35	82.88	77.58	77.18	75.97	76.33	80.75	79.72	79.67	76.66	76.94
En (mol%)	0.39	0.39	0.42	0.43	0.42	0.38	0.39	0.39	0.40	0.43	0.44	0.46	0.40	0.39	0.38	0.39	0.43	0.43	0.42	0.38	0.38
Fs (mol%)	0.12	0.12	0.10	0.10	0.10	0.12	0.11	0.12	0.11	0.10	0.11	0.09	0.11	0.11	0.12	0.12	0.10	0.11	0.11	0.12	0.11
Wo (mol%)	0.49	0.49	0.47	0.47	0.48	0.50	0.50	0.49	0.48	0.46	0.45	0.45	0.49	0.50	0.50	0.49	0.47	0.46	0.47	0.50	0.50

(To be continued)

(Supplementary Table III)

Sample No.	ZK4002 (coarse-grained pyroxenite)																	
Mineral No.	Cpx-1			Cpx-2			Cpx-3			Cpx-4			Cpx-5			Cpx-6		
Analysis No.	1	2	3	1	2	3	1	2	3	1	2	3	1	2	3	1	2	3
SiO_2 (wt.%)	50.21	52.72	50.65	49.80	52.56	53.75	51.94	52.91	50.20	50.23	50.91	52.15	51.77	53.93	49.44	48.63	51.73	50.01
TiO_2	1.55	0.56	1.52	2.00	0.70	0.25	0.87	0.85	1.71	1.70	1.29	0.96	0.92	0.24	1.84	2.30	1.01	1.55
Al_2O_3	3.67	2.11	3.75	4.51	2.00	1.08	2.06	2.39	4.47	3.54	2.92	2.43	2.33	1.21	5.29	5.78	2.78	3.79
Cr_2O_3	0.03	0.01	0.00	0.00	0.01	0.06	0.50	0.20	0.11	0.04	0.37	0.33	0.35	0.01	0.13	0.04	0.17	0.00
FeO	7.33	6.66	7.65	7.06	6.67	5.23	5.46	6.18	7.39	7.76	6.52	6.59	6.90	5.94	7.43	8.15	7.06	7.80
MnO	0.17	0.16	0.18	0.08	0.04	0.12	0.12	0.00	0.08	0.11	0.03	0.20	0.10	0.06	0.06	0.13	0.12	0.17
MgO	13.74	15.36	14.68	13.47	14.67	15.07	15.93	15.50	13.30	13.57	14.32	14.59	15.23	15.87	15.06	13.87	15.73	14.51
CaO	22.21	22.47	21.35	22.39	22.78	24.31	21.89	21.29	22.40	21.91	22.36	22.20	22.00	23.51	18.21	19.03	21.16	21.78
Na_2O	0.36	0.31	0.39	0.39	0.23	0.09	0.32	0.36	0.47	0.39	0.38	0.41	0.38	0.10	1.24	1.12	0.37	0.36
K_2O	0.00	0.00	0.00	0.01	0.00	0.00	0.00	0.00	0.00	0.00	0.00	0.01	0.00	0.00	0.16	0.15	0.00	0.00
Total	99.27	100.4	100.2	99.69	99.65	99.94	99.11	99.70	100.1	99.24	99.10	99.88	99.97	100.9	98.86	99.20	100.1	99.96
Si (O = 6)	1.88	1.94	1.88	1.86	1.95	1.98	1.93	1.95	1.87	1.88	1.90	1.93	1.92	1.97	1.85	1.82	1.91	1.86
Ti	0.04	0.02	0.04	0.06	0.02	0.01	0.02	0.02	0.05	0.05	0.04	0.03	0.03	0.01	0.05	0.06	0.03	0.04

(To be continued)

(Supplementary Table III)

Sample No.	ZK4002 (coarse-grained pyroxenite)																	
Mineral No.	Cpx-1			Cpx-2			Cpx-3			Cpx-4			Cpx-5			Cpx-6		
Analysis No.	1	2	3	1	2	3	1	2	3	1	2	3	1	2	3	1	2	3
Al	0.16	0.09	0.16	0.20	0.09	0.05	0.09	0.10	0.20	0.16	0.13	0.11	0.10	0.05	0.23	0.26	0.12	0.17
Cr	0.00	0.00	0.00	0.00	0.00	0.00	0.01	0.01	0.00	0.00	0.01	0.01	0.01	0.00	0.00	0.00	0.00	0.00
Fe^{2+}	0.23	0.21	0.24	0.22	0.21	0.16	0.17	0.19	0.23	0.24	0.20	0.20	0.21	0.18	0.23	0.26	0.22	0.24
Mn	0.01	0.01	0.01	0.00	0.00	0.00	0.00	0.00	0.00	0.00	0.00	0.01	0.00	0.00	0.00	0.00	0.00	0.01
Mg	0.77	0.84	0.81	0.75	0.81	0.83	0.88	0.85	0.74	0.76	0.80	0.80	0.84	0.86	0.84	0.77	0.87	0.81
Ca	0.89	0.89	0.85	0.89	0.90	0.96	0.87	0.84	0.89	0.88	0.90	0.88	0.87	0.92	0.73	0.76	0.84	0.87
Na	0.03	0.02	0.03	0.03	0.02	0.01	0.02	0.03	0.03	0.03	0.03	0.03	0.03	0.01	0.09	0.08	0.03	0.03
K	0.00	0.00	0.00	0.00	0.00	0.00	0.00	0.00	0.00	0.00	0.00	0.00	0.00	0.00	0.01	0.01	0.00	0.00
Σcation	4.01	4.01	4.01	4.00	4.00	3.99	4.01	3.99	4.00	4.00	4.00	4.00	4.01	4.00	4.03	4.03	4.01	4.02
Mg#	76.96	80.42	77.38	77.28	79.67	83.70	83.86	81.71	76.24	75.72	79.64	79.78	79.73	82.64	78.30	75.19	79.88	76.83
En (mol%)	76.96	0.44	0.43	0.40	0.42	0.42	0.46	0.45	0.40	0.40	0.42	0.43	0.44	0.44	0.47	0.43	0.45	0.42
Fs (mol%)	76.96	0.11	0.13	0.12	0.11	0.08	0.09	0.10	0.12	0.13	0.11	0.11	0.11	0.09	0.13	0.14	0.11	0.13
Wo (mol%)	76.96	0.46	0.45	0.48	0.47	0.49	0.45	0.45	0.48	0.47	0.47	0.47	0.45	0.47	0.41	0.43	0.44	0.45

(To be continued)

(Supplementary Table III)

Sample No.	ZK4001 (fine-grained pyroxenite)						Wjl070101 (olivine pyroxenite)				
Mineral No.	Cpx-1	Cpx-2	Cpx-3	Cpx-4	Cpx-5		Cpx-1			Cpx-2	
Analysis No.	1	1	1	1	1	1	2	3	1	2
SiO_2 (wt.%)	50.00	49.53	50.97	49.52	49.90	49.62	50.49	50.46	50.86	49.64
TiO_2	1.91	2.00	1.55	1.83	1.87	2.15	1.83	1.47	1.23	2.05
Al_2O_3	4.19	4.60	3.46	4.70	4.28	4.38	3.74	3.88	3.11	4.06
Cr_2O_3	0.00	0.00	0.00	0.04	0.00	0.06	0.07	0.32	0.00	0.13
FeO	6.63	7.25	6.62	7.15	7.09	7.83	7.45	7.02	7.28	7.77
MnO	0.16	0.14	0.15	0.10	0.22	0.12	0.08	0.12	0.14	0.11
MgO	14.21	13.82	14.62	14.13	14.20	13.79	14.18	14.77	13.68	13.98
CaO	22.53	22.22	22.21	22.37	22.24	21.30	21.25	20.83	22.31	21.09
Na_2O	0.57	0.57	0.54	0.50	0.53	0.50	0.53	0.49	0.56	0.45
K_2O	0.00	0.00	0.00	0.00	0.00	0.00	0.00	0.00	0.00	0.00
Total	100.2	100.1	100.1	100.3	100.3	99.75	99.62	99.36	99.17	99.29
Si (O = 6)	1.85	1.84	1.89	1.84	1.85	1.85	1.88	1.88	1.91	1.86
Ti	0.05	0.06	0.04	0.05	0.05	0.06	0.05	0.04	0.03	0.06

(To be continued)

(Supplementary Table III)

Sample No.	ZK4001 (fine-grained pyroxenite)					Wj070101 (olivine pyroxenite)				
Mineral No.	Cpx-1	Cpx-2	Cpx-3	Cpx-4	Cpx-5	Cpx-1			Cpx-2	
Analysis No.	1	1	1	1	1	1	2	3	1	2
Al	0.18	0.20	0.15	0.21	0.19	0.19	0.16	0.17	0.14	0.18
Cr	0.00	0.00	0.00	0.00	0.00	0.00	0.00	0.01	0.00	0.00
Fe^{2+}	0.21	0.23	0.20	0.22	0.22	0.24	0.23	0.22	0.23	0.24
Mn	0.01	0.00	0.00	0.00	0.01	0.00	0.00	0.00	0.00	0.00
Mg	0.79	0.77	0.81	0.78	0.78	0.77	0.79	0.82	0.76	0.78
Ca	0.89	0.89	0.88	0.89	0.88	0.85	0.85	0.83	0.90	0.85
Na	0.04	0.04	0.04	0.04	0.04	0.04	0.04	0.04	0.04	0.03
K	0.00	0.00	0.00	0.00	0.00	0.00	0.00	0.00	0.00	0.00
Σ cation	4.02	4.02	4.01	4.03	4.02	4.01	4.00	4.01	4.01	4.01
Mg#	79.26	77.26	79.75	77.89	78.13	75.83	77.24	78.93	77.00	76.22
En (mol%)	0.42	0.41	0.43	0.41	0.42	0.41	0.42	0.44	0.40	0.42
Fs (mol%)	0.11	0.12	0.11	0.12	0.12	0.13	0.12	0.12	0.12	0.13
Wo (mol%)	0.47	0.47	0.47	0.47	0.47	0.46	0.45	0.44	0.47	0.45

Mg# = molar 100 × Mg/(Mg + Fe); En = enstatite, Fs = ferrosilite, Wo = wollastonite.

Supplementary Table IV Chemical compositions of magnetite in the Wajilitag deposit

Sample No.	Wjl070101 (olivine pyroxenite)						Wjl070105 (gabbro)						ZK4001 (fine-grained pyroxenite)							
Mineral No.	Mt-1	Mt-2	Mt-3	Mt-1	Mt-2	Mt-3	Mt-4	Mt-5	Mt-6	Mt-7	Mt-1	Mt-2	Mt-3	Mt-4	Mt-5	Mt-6	Mt-7	Mt-8
Host mineral	Ol	Ol	—	—	Cpx	Cpx	Cpx	—	—	Cpx	—	—	—	—	—	—	—	—
Analysis No.	1	1	1	1	1	1	1	1	1	1	1	1	1	1	1	1	1	1
TiO$_2$ (wt.%)	9.02	7.26	1.44	10.48	18.19	9.79	7.46	8.49	8.76	5.58	12.44	17.86	12.14	12.07	10.98	17.76	9.08	14.44
Al$_2$O$_3$	2.06	2.13	1.95	0.53	2.21	0.96	1.58	1.90	1.95	0.71	3.25	2.87	2.75	3.92	3.62	2.99	2.86	2.03
Cr$_2$O$_3$	1.80	1.94	0.07	0.52	0.21	3.15	0.55	1.14	1.12	0.68	0.06	0.02	0.09	0.03	0.06	0.04	0.00	0.12
FeO$_T$	80.97	82.89	89.38	82.33	72.14	80.99	84.69	81.54	82.44	87.95	77.50	71.86	78.51	77.28	79.23	71.95	81.55	77.10
Fe$_2$O$_3$	47.07	50.48	63.85	47.58	34.16	45.86	53.30	49.23	49.81	57.40	43.03	33.12	44.09	43.24	45.40	32.58	48.36	38.86
FeO	38.61	37.47	31.92	39.51	41.39	39.73	36.72	37.24	37.62	36.30	38.78	42.06	38.84	38.38	38.38	42.63	38.03	42.13
MnO	0.37	0.20	0.08	0.48	0.43	0.66	0.46	0.34	0.35	0.25	0.38	0.50	0.37	0.37	0.39	0.39	0.36	0.30
MgO	0.40	0.29	0.40	0.49	4.43	0.00	0.88	0.96	1.18	0.07	2.74	3.64	2.49	2.91	2.23	3.20	0.96	1.41
Total	99.33	99.76	99.71	99.60	101.0	100.1	100.9	99.29	100.8	101.0	100.7	100.1	100.8	100.9	101.1	99.59	99.65	99.29
Ti (O = 4)	0.26	0.21	0.04	0.30	0.49	0.28	0.21	0.24	0.24	0.16	0.34	0.49	0.33	0.33	0.30	0.49	0.26	0.41
Al	0.09	0.09	0.09	0.02	0.09	0.04	0.07	0.08	0.09	0.03	0.14	0.12	0.12	0.17	0.16	0.13	0.13	0.09
Cr	0.05	0.06	0.00	0.02	0.01	0.09	0.02	0.03	0.03	0.02	0.00	0.00	0.00	0.00	0.00	0.00	0.00	0.00
Fe^{3+}	1.34	1.43	1.83	1.36	0.92	1.31	1.50	1.40	1.39	1.63	1.18	0.90	1.21	1.18	1.24	0.90	1.36	1.09
Fe^{2+}	1.22	1.18	1.02	1.26	1.24	1.26	1.15	1.18	1.17	1.15	1.18	1.27	1.19	1.16	1.17	1.30	1.19	1.32
Mn	0.01	0.01	0.00	0.02	0.01	0.02	0.01	0.01	0.01	0.01	0.01	0.02	0.01	0.01	0.01	0.01	0.01	0.01
Mg	0.02	0.02	0.02	0.03	0.24	0.00	0.05	0.05	0.07	0.00	0.15	0.20	0.14	0.16	0.12	0.17	0.05	0.08
Σ cation	3.00	3.00	3.00	3.00	3.00	3.00	3.00	3.00	3.00	3.00	3.00	3.00	3.00	3.00	3.00	3.00	3.00	3.00

(To be continued)

(Supplementary Table IV)

Sample No.									ZK4002 (coarse-grained pyroxenite)											
Mineral No.	Mt-1			Mt-2				Mt-3		Mt-4			Mt-5		Mt-6		Mt-7			
Host mineral	Cpx			Cpx				Cpx		Hbl			Ol		Ol		Ol			
Analysis No.	1	2	3	1	2	3	4	1	2	1	2	3	1	2	1	2	1	2		
TiO$_2$ (wt.%)	7.69	7.85	6.67	8.72	6.89	7.22	7.10	10.40	10.38	5.15	4.25	4.65	9.17	9.36	6.66	6.89	10.30	8.63		
Al$_2$O$_3$	2.54	3.39	2.67	4.44	3.35	3.50	3.42	3.61	3.59	0.70	2.85	0.55	3.03	3.39	3.85	4.21	3.44	3.36		
Cr$_2$O$_3$	0.21	0.26	0.24	0.33	0.26	0.27	0.24	0.32	0.25	1.20	1.04	1.11	0.32	0.33	0.29	0.36	1.01	1.52		
FeO$_T$	83.67	82.67	84.93	78.81	83.77	82.41	83.13	80.20	79.41	87.27	86.34	86.98	81.29	80.28	82.25	81.80	78.40	79.96		
Fe$_2$O$_3$	51.93	50.79	53.91	46.75	52.80	51.32	52.05	46.02	45.31	57.05	56.96	57.84	48.04	47.28	51.85	50.77	44.01	47.18		
FeO	36.95	36.96	36.42	36.75	36.26	36.23	36.30	38.79	38.64	35.94	35.09	34.94	38.06	37.74	35.59	36.12	38.80	37.51		
MnO	0.23	0.20	0.18	0.27	0.27	0.22	0.15	0.33	0.33	0.04	0.16	0.22	0.23	0.43	0.23	0.22	0.48	0.35		
MgO	1.07	1.29	0.91	1.67	1.18	1.25	1.26	1.67	1.53	0.00	0.29	0.07	1.18	1.38	1.30	1.14	1.09	1.10		
Total	100.6	100.7	101.0	98.91	101.0	100.0	100.5	101.1	100.0	100.1	100.6	99.38	100.0	99.90	99.77	99.71	99.13	99.65		
Ti (O = 4)	0.22	0.19	0.19	0.24	0.19	0.20	0.20	0.29	0.29	0.15	0.12	0.13	0.26	0.26	0.19	0.19	0.29	0.24		
Al	0.11	0.15	0.12	0.19	0.15	0.15	0.15	0.16	0.16	0.03	0.13	0.02	0.13	0.15	0.17	0.18	0.15	0.15		
Cr	0.01	0.01	0.01	0.01	0.01	0.01	0.01	0.01	0.01	0.04	0.03	0.03	0.01	0.01	0.01	0.01	0.03	0.04		
Fe^{3+}	1.45	1.41	1.50	1.31	1.46	1.44	1.45	1.26	1.26	1.64	1.60	1.67	1.34	1.32	1.45	1.42	1.24	1.32		
Fe^{2+}	1.15	1.14	1.13	1.14	1.12	1.13	1.12	1.18	1.19	1.15	1.10	1.12	1.18	1.17	1.11	1.12	1.21	1.17		
Mn	0.01	0.01	0.01	0.01	0.01	0.01	0.00	0.01	0.01	0.00	0.01	0.01	0.01	0.01	0.01	0.01	0.02	0.01		
Mg	0.06	0.07	0.05	0.09	0.06	0.07	0.07	0.09	0.08	0.00	0.02	0.00	0.07	0.08	0.07	0.06	0.06	0.06		
Σ cation	3.00	3.00	3.00	3.00	3.00	3.00	3.00	3.00	3.00	3.00	3.00	3.00	3.00	3.00	3.00	3.00	3.00	3.00		

Ol = olivine, Cpx = clinopyroxene, Hbl = hornblende; "–" means the magnetite interstitial to silicate minerals.

Supplementary Table V Chemical compositions of ilmenite in the Wajilitag deposit

Sample No.	Wjl070101 (olivine pyroxenite)		Wjl070105 (gabbro)			ZK4001 (fine-grained pyroxenite)				
Mineral No.	Ilm-1	Ilm-2	Ilm-1	Ilm-2	Ilm-1	Ilm-2	Ilm-3	Ilm-4	Ilm-5	Ilm-6
Host mineral	Ol	Ol	–	Cpx	–	–	–	–	–	–
Analysis No.	1	1	1	1	1	1	1	1	1	1
TiO_2 (wt.%)	49.74	50.20	52.00	51.15	53.88	51.61	53.00	52.75	53.78	52.54
Al_2O_3	0.08	0.06	0.06	0.01	0.06	0.06	0.09	0.05	0.04	0.03
Cr_2O_3	0.13	0.12	0.06	0.00	0.06	0.06	0.03	0.01	0.03	0.02
FeO_T	47.75	47.76	42.11	46.43	38.58	40.48	39.00	39.74	40.32	41.96
Fe_2O_3	6.64	6.13	4.53	3.45	3.72	6.74	5.54	4.57	3.43	5.21
FeO	41.78	42.25	38.04	43.33	35.23	34.42	34.01	35.62	37.23	37.27
MnO	1.61	1.36	1.08	1.62	0.50	0.48	0.54	0.49	0.67	0.55
MgO	0.48	0.63	4.12	0.55	6.98	6.30	7.18	6.25	5.77	5.21
Total	100.5	100.7	99.89	100.1	100.4	99.67	100.4	99.75	101.0	100.8
Ti (O = 3)	0.94	0.94	0.96	0.97	0.97	0.94	0.95	0.96	0.97	0.95
Al	0.00	0.00	0.00	0.00	0.00	0.00	0.00	0.00	0.00	0.00
Cr	0.00	0.00	0.00	0.00	0.00	0.00	0.00	0.00	0.00	0.00
Fe^{3+}	0.13	0.12	0.08	0.07	0.07	0.12	0.10	0.08	0.06	0.09
Fe^{2+}	0.88	0.88	0.78	0.91	0.70	0.70	0.68	0.72	0.75	0.75
Mn	0.03	0.03	0.02	0.03	0.01	0.01	0.01	0.01	0.01	0.01
Mg	0.02	0.02	0.15	0.02	0.25	0.23	0.26	0.23	0.21	0.19
Σ cation	2.00	2.00	2.00	2.00	2.00	2.00	2.00	2.00	2.00	2.00

(To be continued)

(Supplementary Table V)

Sample No.	ZK4002 (coarse-grained pyroxenite)									
Mineral No.	Ilm-1		Ilm-2				Ilm-3		Ilm-4	
Host mineral	Cpx		Cpx				Cpx		Ol	
Analysis No.	1	2	1	2	3	4	1	2	1	2
TiO_2 (wt.%)	52.20	52.08	52.50	53.33	52.78	52.19	53.15	52.59	51.63	52.07
Al_2O_3	0.03	0.04	0.03	0.00	0.03	0.02	0.00	0.01	0.05	0.07
Cr_2O_3	0.00	0.00	0.00	0.03	0.00	0.00	0.00	0.00	0.00	0.00
FeO_T	42.99	42.93	41.74	42.42	43.47	42.95	41.77	42.53	43.70	43.44
Fe_2O_3	4.62	4.31	2.19	2.01	3.96	4.00	1.82	2.99	4.47	3.99
FeO	38.84	39.05	39.76	40.61	39.91	39.36	40.13	39.83	39.67	39.85
MnO	0.46	0.42	0.71	0.59	0.61	0.46	0.71	0.64	0.73	0.75
MgO	4.23	4.06	3.72	3.75	3.85	3.95	3.89	3.80	3.29	3.38
Total	100.4	99.97	98.92	100.3	101.1	99.97	99.70	99.86	99.85	100.1
Ti (O = 3)	0.96	0.96	0.98	0.98	0.96	0.96	0.98	0.97	0.96	0.96
Al	0.00	0.00	0.00	0.00	0.00	0.00	0.00	0.00	0.00	0.00
Cr	0.00	0.00	0.00	0.00	0.00	0.00	0.00	0.00	0.00	0.00
Fe^{3+}	0.08	0.08	0.04	0.04	0.07	0.07	0.03	0.06	0.08	0.07
Fe^{2+}	0.79	0.80	0.83	0.83	0.81	0.81	0.83	0.82	0.82	0.82
Mn	0.01	0.01	0.01	0.01	0.01	0.01	0.01	0.01	0.02	0.02
Mg	0.15	0.15	0.14	0.14	0.14	0.14	0.14	0.14	0.12	0.12
Σ cation	2.00	2.00	2.00	2.00	2.00	2.00	2.00	2.00	2.00	2.00

Ol = olivine, Cpx = clinopyroxene; "—" means the ilmenite interstitial to silicate minerals.

Supplementary Table VI Chemical compositions of the coexisting magnetite–ilmenite in the Wajilitag deposit

Sample No.	Wjl070101 (olivine pyroxenite)				Wjl070105 (gabbro)				ZK4001 (fine-grained pyroxenite)				ZK4002 (coarse-grained pyroxenite)				
Mineral No.	Mt-1	Ilm-1	Mt-2	Ilm-2	Mt-6	Ilm-1	Mt-7	Ilm-2	Mt-1	Ilm-1	Mt-4	Ilm-3	Mt-1	Ilm-3	Mt-2	Ilm-2	Mt-6	Ol	Ilm-4
Host mineral	Ol		Ol		—		Cpx		—		—		Cpx		Cpx		Ol	
TiO_2 (wt.%)	9.02	49.74	7.26	50.20	8.76	52.00	5.58	51.15	12.44	53.88	12.07	53.00	7.40	52.87	7.48	52.70	6.77	51.85
Al_2O_3	2.06	0.08	2.13	0.06	1.95	0.06	0.71	0.01	3.25	0.06	3.92	0.09	2.86	0.01	3.68	0.02	4.03	0.06
Cr_2O_3	1.80	0.13	1.94	0.12	1.12	0.06	0.68	0.00	0.06	0.06	0.03	0.03	0.24	0.00	0.27	0.01	0.33	0.00
FeO_T	80.97	47.75	82.89	47.76	82.44	42.11	87.95	46.43	77.50	38.58	77.28	39.00	83.76	42.15	82.03	42.64	82.02	43.57
Fe_2O_3	47.07	6.64	50.48	6.13	49.81	4.53	57.40	3.45	43.03	3.72	43.24	5.54	52.21	2.41	50.73	3.04	51.31	4.23
FeO	38.61	41.78	37.47	42.25	37.62	38.04	36.30	43.33	38.78	35.23	38.38	34.01	36.78	39.98	36.38	39.91	35.85	39.76
MnO	0.37	1.61	0.20	1.36	0.35	1.08	0.25	1.62	0.38	0.50	0.37	0.54	0.21	0.67	0.23	0.59	0.23	0.74
MgO	0.40	0.48	0.29	0.63	1.18	4.12	0.07	0.55	2.74	6.98	2.91	7.18	1.09	3.85	1.34	3.82	1.22	3.34
Total	99.33	100.5	99.76	100.7	100.8	99.89	101.0	100.1	100.7	100.4	100.9	100.4	100.8	99.78	100.1	100.1	99.74	99.98
Ti	0.26	0.94	0.21	0.94	0.24	0.96	0.16	0.97	0.34	0.97	0.33	0.95	0.21	0.98	0.21	0.97	0.19	0.96
Al	0.09	0.00	0.09	0.00	0.09	0.00	0.03	0.00	0.14	0.00	0.17	0.00	0.13	0.00	0.16	0.00	0.18	0.00
Cr	0.05	0.00	0.06	0.00	0.03	0.00	0.02	0.00	0.00	0.00	0.00	0.00	0.01	0.00	0.01	0.00	0.01	0.00
Fe^{3+}	1.34	0.13	1.43	0.12	1.39	0.08	1.63	0.07	1.18	0.07	1.18	0.10	1.46	0.04	1.41	0.06	1.44	0.08
Fe^{2+}	1.22	0.88	1.18	0.88	1.17	0.78	1.15	0.91	1.18	0.70	1.16	0.68	1.14	0.82	1.13	0.82	1.11	0.82
Mn	0.01	0.03	0.01	0.03	0.01	0.02	0.01	0.03	0.01	0.01	0.01	0.01	0.01	0.01	0.01	0.01	0.01	0.02
Mg	0.02	0.02	0.02	0.02	0.07	0.15	0.00	0.02	0.15	0.25	0.16	0.26	0.06	0.14	0.07	0.14	0.07	0.12
Σ cation	3.00	2.00	3.00	2.00	3.00	2.00	3.00	2.00	3.00	2.00	3.00	2.00	3.00	2.00	3.00	2.00	3.00	2.00
Σ O	4.00	3.00	4.00	3.00	4.00	3.00	4.00	3.00	4.00	3.00	4.00	3.00	4.00	3.00	4.00	3.00	4.00	3.00

Ol = olivine, Cpx = clinopyroxene; "–" means the magnetite–ilmenite intergrowths interstitial to silicate minerals.

Index

A
Altyn-Tagh orogen 1,5

C
Continental flood basaltic lavas 27
Crustal contamination 75,78,80,82-85,87-91,170-171
Crystalline basement 2,28
Cu–Ni–PGE mineralization 153,166, 174-175

D
Depleted mantle component 75
Domal uplift 49,109-110
Double-layer structure 3
Dike swarm 9,11-12,17,27,48-49,53,65, 109-110,116-118,136

E
Enriched mantle component 75,91

F
Fe–Ti–V oxide deposit 53,153-154
Field contact relationship 27,31,37,62

G
Geochemical comparison 81,109,120
Geochemical features 17,75,82,89, 129,139,142-144

Geodynamic model 109,135,137-138,142

I
Intermediate-felsic volcanic rock 17
Intrusive rock 10,27-28,33,37,45-46,50,57-58,60,63,75-76,78,82-84,91,162
Isotopic characteristics 75,79,82-83, 91,141

K
Kaipaizileike Formation 27-29,38-43,50-52,57,63-65,105
Kunlun orogen 1
Kupukuziman Formation 14,30,38-44,49,51,57,63-65,80,105
Kuqa Depression 1,7-8

L
LA-ICP-MS 53-55,61,63,65

M
Magma evolution 75,87
Magmatic process 153,156,166,173
Main rock units 27,75,156
Mantle plume model 109
Metallogenesis 154
Metamorphic basement 3

P
Permian igneous rock 37,47,57,109,111
Phanerozoic tectonostratigraphy 6
Plume–lithosphere interaction 75,91

R
Research history 13

S
SCLM 18,91,136-137,141,143
SHRIMP 15,53-54,56-58,60,63-65,85
Silicate crystals 153,161
Source isotopic heterogeneity 75,90-91
Spatial distribution 3,27,29,37,43,45, 50-51,65,111
Sr–Nd–Hf isotopes 75,84

T
Tectonic evolution 1,3
Tectonic unit 1,2
Temporal order 27,52
Tempo-spatial feature 27-64,109,136, 139,142-143
Three group basalts 167,168,170-173
Tianshan orogen 1,5,7,85-86,90
Two-stage melting model 109,136

W
Wajilitag complex 154-156,160-164